REDWOOD
CLASSIC

REDWOOD CLASSIC

by

Ralph W. Andrews

(Opposite) "AVENUES OF POETIC GRANDEUR" said John Muir of California's heritage. The naturalist struggled hard to preserve the trees for beauty alone. (Ericson's photo from California Redwood Association.)

Schiffer Publishing Ltd®

4880 Lower Valley Road, Atglen, Pennsylvania 19310

Schiffer Books are available at special discounts for bulk
purchases for sales promotions or premiums. Special
editions, including personalized covers, corporate imprints,
and excerpts can be created in large quantities for special
needs. For more information contact the publisher:

Published by Schiffer Publishing Ltd.
4880 Lower Valley Road
Atglen, PA 19310
Phone: (610) 593-1777; Fax: (610) 593-2002
E-mail: Info@schifferbooks.com

For the largest selection of fine reference books on this and
related subjects, please visit our web site at:
www.schifferbooks.com
We are always looking for people to write books on new
and related subjects. If you have an idea for a
book please contact us at the above address.

This book may be purchased from the publisher.
Include $5.00 for shipping.
Please try your bookstore first.
You may write for a free catalog.

In Europe, Schiffer books are distributed by
Bushwood Books
6 Marksbury Ave.
Kew Gardens
Surrey TW9 4JF England
Phone: 44 (0) 20 8392-8585; Fax: 44 (0) 20 8392-9876
E-mail: info@bushwoodbooks.co.uk
Website: www.bushwoodbooks.co.uk
Free postage in the U.K., Europe; air mail at cost.

DEDICATED

to the courage

and ingenuity of

The Man With The Axe

FOREWORD

Anyone leafing through this book will quickly see it is not a history of the redwoods, not a eulogy on their poetic significance, not a technical treatise of the tree and wood. A closer look will show it to be simply a reportorial effort to bring to present understanding the forgotten struggles and triumphs of the people to whom the redwoods meant life and a living in a new land.

Is there a better means of achieving this than by recreating early scenes with photographs of the day supplemented by a framework of documented stories and statements?

The experience of this author and his publisher in producing several "This Was" books has established a general pattern of presentation which the public, critics, and book sellers have found lively and a needed addition to the lore of the Western Shore.

The publisher's success with this pattern has led to his specializing in this type of book, hesitatingly termed the "picture book" or "historical pictorial." People who are not congenital readers—and they are legion—can better visualize yesterday a hundred years ago by studying a photograph than by struggling through those vaunted ten thousand words. The only requirement is that the photograph tell the story.

Redwood Classic follows the pattern of the author's foregoing books. Its preparation starts with photographs and is limited to them. All paintings, sketches and other fanciful interpretations have been carefully avoided and extensive research carried on to find authentic and story-telling photographs taken by people who saw the redwoods fall. To give further verisimilitude original writing has been held to a minimum since this author's conceptions are no more accurate or valuable than another's today. Instead pertinent writings of the period are used, chosen for their humanities and information rather than their qualities of style and expression. The book so becomes an historical display between covers. And if it does not have the sights, sounds, smells and color of the redwoods past the author is as disappointed as the reader.

A technical point should be mentioned here. Many books of this pictorial nature reproduce good and bad photographs by lithography, a less costly and inferior method as to printed quality and paper stock. The Superior Publishing Company has never compromised with this tendency, has always used the fine-screen copper halftone printed on high quality enameled paper. This principle has been highly appreciated by the public and trade, has attracted authors who have justifiable pride in the reproduction of their work.

The task of digging into old records and musty bins is always made easier by the esprit de corps of historians and historical societies, the eager cooperation of individual collectors and lumber firms. People ask: "How do you find these pictures?" And the answer is: "Like gold—wherever it is."

The most help came from such treasure houses as the Bancroft Library, Forestry Library and the Agriculture Extension Division of the University of California at Berkeley; the Tulare County Historical Society through its dedicated "Los Tulares" editor, Harold G. Schutt; the Oakland Public Library; Wells Fargo Bank History Room; California Redwood Association of San Francisco; Union Lumber Company; Pacific Lumber Company and Rockport Redwood Company.

Through a great stroke of luck Curtis Annand of Bremerton, Washington, came forward with the many negatives of photographer C. C. Curtis, his grandfather, and through him valuable data was secured from his mother and Curtis' daughter, Mrs. Cecile Pennell of Santa Cruz. Another important discovery was Nannie Escola, pioneer and collector of Mendocino who cooperated wholeheartedly.

The publications from which material has been quoted are mentioned in each instance and include Humboldt Times, Del Norte Triplicate (Crescent City, Mendocino Coast, McNairn and MacMullen's Ships Of The Redwood Coast. Other publications used for reference include Alfred Powers' Redwood Country, Joseph Henry Jackson's collection in Western Gate, The Western Railroader of San Mateo and the file of Hammond-California Lumber Company's Redwood Log. To all the people and firms responsible the author extends public thanks.

RALPH W. ANDREWS

CONTENTS

AND THERE WERE NO STUMPS

The first white men who saw the California redwoods were hungry. Beauty was beauty but they were practical men, even the Franciscan fathers. The great *palo colorados* were profoundly impressive, and the very air about them hallowed by their eminence, yet they had little more utility than to provide cover for marauding Indians. There was only one thing to do with them they were sure — cut as many as necessary to open up the land for farms on which to raise food.

There seemed to be no end to the redwood forest. On the coast it extended southward from the Chetco River in southwestern Oregon to Salmon Creek near Monterey. The belt was broken and variable in width but most of the trees grew within twenty miles of the coastline. And in the Sierra Nevada's Kings River area, a still larger *sequoia gigantea* grew profusely.

The fact that so few of these giant trees were felled by the early settlers was due to this being a tremendous task. Of what use was an axe and a burro when you had in front of you a tree trunk twenty feet in diameter, over two hundred feet high which might weigh four hundred tons? These men could simply not conceive of any method to take down and work up the big redwoods.

(*opposite*) **FALLING BIG SEQUOIA GIGANTEA** at Mountain Home about 1904. Chopper at right is early settler Earl McDonald. (Photo Harold G. Schutt Collection)

"AND TREES SO BIG 20 MEN with arms outstretched can't reach around them." Unbelievable were the tales of redwoods sent back East in 1850 - 1860. This was the General Grant Tree. (Curtis photo from Annand Collection)

CANES AND SUNDAY BEST at the Fallen Monarch and General Stoneman sequoia. Tulare County pioneers went to the redwoods to ride horseback through this old log. (Curtis photo from Annand Collection)

A few of the first Californians of course—the naturalists and the sages — were not of this practical breed. They saw the splendid "monarchs of antiquity" as pillars of temples, as giant columns which offered an infinite variety of expressions and which were possessed of spiritual qualities and of mysterious influence. In the Century Magazine of November 1891, John Muir wrote:

"At first sight it would seem that these mighty granite temples could be injured but little by anything that man may do. But it is surprising to find how much of our impressions in such cases depends upon the delicate bloom of the scenery, which in the more accessible places, is so easily rubbed off.

"I saw the King's River Valley in its midsummer glory sixteen years ago, when it was wild and when the divinely balanced beauty of the trees and flowers seemed to be reflected and doubled by all the outlooking rocks and streams as though they were mirrors, while they in turn were mirrored in every garden and grove. In that year (1875) I saw the following notice on a tree in the King's River Yosemite:

We, the undersigned, claim this valley
for the purpose of raising stock.
and three names were signed thereto.

"And I feared this vegetation would soon perish. This spring I made my fourth visit to this valley to see what damage had been done and to inspect the forests. At the new King's River mills we found the *sequoia* giants as well as the pines and firs were being ruthlessly turned into lumber. Sixteen years ago I saw five mills on or near the *sequoia* belt all of which were cutting more or less of the big tree lumber. Now I am told the number of mills along the belt in the basin of the King's, Kaweah and Tule Rivers is doubled, the capacity more than doubled. Scaffolds are built around the great brown shafts above the swell of the base and several men armed with long saws and axes gnaw and wedge them down with damnable industry. The logs found to be too large are blasted to manageable dimensions with powder. It seems incredible that the Government should have abandoned so much of the forest's cover of the mountains to destruction. As well sell the rain clouds and the snow and the rivers, to be cut up and carried away, if that were possible. Surely it is high time that something should be done to stop the extension of the present barbarous, indiscriminating method of harvesting the lumber crop."

Very little was done because there were too

PRIDE OF PETRIFIED FOREST—Sonoma County's 20-acre showplace brought visitors from near and far in California's early days. This section was 68 feet long, 11 feet in diameter. (Cherry photo from California Redwood Assn.)

"Although to a thoughtful man," said Benjamin F. Taylor, "the petrified trees are the most impressive things in California. They overwhelm your vanity with gray cairns of what once danced in the rain, whispered in the wind, blossomed in the sun . . . What a rocking of the cradle there must have been when the earth quaked and lava put these trees in flinty armor and transfused their veins with dumbness."

Poet Taylor was referring to Sonoma County's "stone forest" which he visited in 1878. He found the land claim had been taken by a shorebound sailor named C. Evans who was selling chips of trees and charging half a dollar admission to the area.

The petrified boles of the three hundred trees are spread over the twenty acres, many of them with thick bark intact, some showing blackened signs of char, some even harboring specimens of worms. The ring count of the largest fossil, a hundred and twenty-five feet long, indicated a thousand years of growth when it fell.

many practical men who saw the redwoods as a barrier to progress and the wood a merchantable product. They were men like Frank Maryatt who came up to the Russian River country from the Sonora gold hills in 1850 and wrote in his book "Mountains And Molehills" in 1855:

"Thomas and I proceeded in search of a backwoodman's hut which we had been informed existed in this direction; after following the river for some time we ascended a steep hill, from the summit of which was presented the most lovely panorama — beneath us the thickly wooded plain extended for miles — on one side bounded by mountains, on the other, melting away in the hazy blue distance, the windings of the Russian River were marked distinctly in contrast to the dark rocks and foliage that lined its banks, while immediately beneath us was a forest of firs and redwood trees over which vultures wheeled incessantly and not even the sound of an insect disturbed the silence of the scene.

"From this hill we discovered the hut of which we were in search, situated near a running stream and surrounded by towering redwood trees. We found the occupant at home; he was a tall, sinewy man, a Missourian by the name of March, and he at once cheerfully assisted us. He lent us his mule to bring up our baggage and by nightfall we were

REDWOOD CITY ABOUT 1860—Lumber loading waterfront near present Broadway. (Photo courtesy San Mateo County Historical Association)

encamped within a few yards of his hut. There were two other backswoodsmen living with March and these three had just completed unaided a sawmill to which they had applied the power of the stream by means of an overshot wheel. The heavy beams that formed the mill frame, the dam and race, had all been constructed from the adjacent forest trees, and now that the work was completed, wanting only the saw, for which they intended to go to San Francisco, it seemed incredible that so large a frame could be put together by so small a number of men. This sawmill erected in the forest and of the forest, raising its long beams from the midst of the romantic scenery that surrounded it, was a glorious instance of what energy will accomplish, and of the rapidity with which each man in an American colony contributes toward the development of the new country's resources.

"And it contrasts strangely with the languid inertness of these communities, who with equal brains and hands, ponder and dream over the means of supplying its wants, even when they have long been felt; to see that here even the educated backwoodsman devotes his time and energy to preparing for the wants to come; buoyed up by an admirable confidence in the rapid growth and prosperity of his country, which confidence is part of his education, and one great secret of his success. If the Americans go ahead, it is principally because they look ahead. March, when he planned his mill and felled his first tree in this solitary forest, ranked with those who wrote from the tents of San Francisco for steam engines and foundries. Now as I write, these latter are performing their daily work in the city and have become essential to its wants whilst March's mill, seemingly so out of place when I first saw it, can now barely supply the wants of the numerous agricultural population that is settling around it. March and his companions lived entirely on game, which he assured me abounded; and as for the present at all events I could not proceed, I determined at his advice to walk over the hills and look at a valley on which he strongly recommended me to 'squat,' we therefore started the next morning in search of it, following the directions that March had given us."

In all parts of the young and restless United States and from many foreign lands, men were hearing about the new country of California and the great, red trees older than any living thing. Lumbering here would be a virgin field that required great ingenuity to enter, much skill and hard work to conquer but which held great rewards for the future. They left their native farms and started for California, men like George Patterson, late of Ireland.

On May 19, 1847, Patterson strode the deck of the English bark *Columbia* and solemnly wondered if his plan to desert was a wise one. He had pondered it for weeks and now that the vessel was quietly rolling with the swells in San Francisco Bay he knew he must make up his mind now. He had grave doubts about the wisdom of entering this new country but would things be any worse than this soul-wracking work aboard ship? Then he saw the tall bulks of the giant redwoods on three sides of him and thought surely there was salvation and pleasant living in the great groves. He deserted.

Patterson quickly found that various tribes of Indians and a few more intrepid Spaniards had made a little use of the redwods. The Russians had been more business like. They had been cutting the big trees since the early 1800s, had utilized the wood to considerable extent and shipped some out.

Fort Ross itself and the buildings within its confines were all built of redwood. Huge logs formed the palisades, blockhouses, warehouses, barracks and chapel, the latter of handsplit boards. Redwood had been used, unsuccessfully because of the nature of the wood and of insufficient skills, tools and methods, in several of their ships. The traffic in redwood for fuel was carried on extensively, boats ferrying many thousands of cords to San Francisco and the Farrallons where the Russians had another colony.

While they strained their backs with the big

logs, a few stalwarts were whipsawing redwood or splitting it by hand. One, a Spaniard named Jose Amesto, sold shingles from his Corralitos Canyon ranch near Monterey in 1830 and ten years later "Scotch" Whalley whipsawed redwood near the same spot. At this time Pedro Sainswain, William Blackburn and Isaac Graham wrestled a living with crude pit and platform sawing near Santa Cruz as James Dawson did in the Russian River country. And there was the McMillan mill on Los Gatos Creek.

These sporadic lumbering efforts were all done by hand and so was George Patterson's when he went into the San Antonio hills east of San Francisco Bay. With William Parker and later Harry J. Bee, he sold lumber to John A. Sutter.

Amateur loggers came from San Francisco, built a road through the redwood forest to the gap in the Piedmont Hills and sold the hand-cut timber to Dr. Robert Semple who shipped it across Suisan Bay to Benecia. Napoleon Bonaparte Smith, his brother Harry Clay Smith and William Mendenhall with Elam Brown of Santa Clara, whipsawed redwood here. Another of these rugged one-man enterprises was that of Jacob Wright Harlan who left a record of a short foray into the timber in a publication, "California '46 To '88":

"On arriving at Mission San Jose, I separated from my comrades, most of whom I have not seen since. I was welcomed at the home of my Uncle Harlan where I remained for a few days. It was necessary to get work and work was not easy to get. Gold had not yet been discovered and there was but little doing, wages being 8 and 10 dollars a month. So I determined to go into the redwood forests on the east side of the San Antonio range of hills, to the eastward of the present site of East Oakland and here try to make shingles.

"WOOD THAT NEVER ROTS" the pioneers called redwood. This log was still sound with over a thousand years of growth around it when photographed in Dolbeer and Carson's woods, Humboldt County, about 1908. (Photo courtesy E. J. Stewart.)

UNDERCUT TOOK DAY AND A HALF, sawing and falling two more when this redwood was downed at Northspur camp on Noyo River. Choppers Matt Mantilla (left with axe) and Emil Johnson (right with axe) made falling bed out of six 3-foot logs and cleared away chips so butt would rest on ground. Henry Gordon (sitting on log) marked lengths for buckers at 28 cents an hour. Other men identified: between choppers, Bee Charlie Bettigo; on stump Swanti Maki and Bill Turner. (Frederick photo from Escola Collection.)

STARTING AN EIGHT-DAY JOB—The imagination can hardly grasp the audacity of man to attack these great bulks with little axes. (C. C. Curtis photo from Annand Collection)

"I was entirely unacquainted with this kind of work yet could only try and so hired Richard Swift and went to work. We cut down a very large redwood and worked it up into 15 thousand shingles which occupied us a month. I had them delivered to San Antonio Landing, now East Oakland, and shipped on a flatboat to San Francisco where they were sold to W. A. Leidesdorf for $5 a thousand. I paid Swift his wages and all expenses and had $50 left.

"At this time there were four stores or principal warehouses in San Francisco. The town had begun to grow a little, its population might be 300. Lumber was scarce and not easy to get. It all had to be sawed by whip saws as there were no sawmills in the country.

"After disposing of the shingles, Leidesdorf asked me if I could take a contract to fence in 1650 vara lots in San Francisco belonging to Commodore Sloat, C. Stockton, Col. Fremont and some others. The fence was to consist of two rails with mortised posts and space 3 feet between rails. He said their object was to prevent squatters from occupying the lots. I thought well of the proposition and we entered into an agreement in which there was no time limit mentioned for the completion of the work.

"I went to San Jose and found David Williamson who was working in Oliver's grist mill and earning $2.50 a day. I showed him the contract and told him what I had done and asked him to be a partner in the work. At first he hesitated, saying he was in good employment, earning good wages and that my job might be a failure. I said, 'All right, David. We will see,' and I mounted my horse and rode off, but presently he called me back and said he would join me.

"We then hired Swift and went into the redwoods, where we cut down trees and split them into posts and rails. We bought two yoke of oxen and a wagon and hired my cousin Joel Harlan to help me haul the stuff to San Antonio for shipment to San Francisco. While we were hauling, Williamson was in the city mortising the posts and as soon as we got them all over to the city, I went there with him to finish the work. There was then no way to take wagons and teams across the bay from what is now Oakland to San Francisco except by the way of San Jose.

"We had our tent on the sandhills on Market Street on a lot where the Palace Hotel now stands.

126 PEOPLE AND 30 MUSTACHES on one redwood stump. (Curtis photo from Bancroft Library, University of Calif.)

It was nearly all sand hills about there at the time with scrub oak bushes all over the neighborhood. We had no tea or coffee but yerba buena (a creeping vine) grew in plenty under the bushes. We made tea of it and drank nothing else. I believe it is more wholesome than Chinese coffee or tea.

"We began this work on July 6 and finished September 20, 1847 to the satisfaction of Leidesdorf who paid as agreed. After paying all expenses we had $500 to divide and went to camp, sat down on blankets on the deep sand and made the division. David declared it the best strike he had ever made and was going back to Cincinnati. I tried to get him to stay but he was set on going back and did so."

Now the power sawmills were springing up, as word went abroad that the redwoods were fair game and being taken. In 1841 two water-power mills started sawing, the one of Nathaniel Speir and Capt. William S. Hinckley in the Oakland Hills, the other near Mount Hermon in the Santa Cruz country which had been built by Peter Lassen. Each of these areas has been termed the "cradle of the California lumber industry" and each has a legitimate claim.

The fifty mile strip, thirteen miles wide, ending north and east in the Santa Clara Valley and south and west at the Pacific Ocean, was the redwood area first discovered by the Spanish missionaries and saw the earliest power sawmills.

In 1842 Isaac Graham purchased the Lassen mill on Zayante Creek for one hundred mules while Pedro Sainswain set one up on the Canada del Rincon below Powdermill Flat, near the present site of Paradise Park, Santa Clara. Near Ben Lomond, on Love Creek, was the sawmill of Capt. Harry Love who later led the posse that captured Joaquin Murietta.

On Soquel Creek, a small mill was operated in 1845 by John Hames and John Raubenbiss and near Redwood City, "Bill, The Sawyer" Smith and James Peace were whipsawing. In 1847 the Mountain Home mill was started by Coppinger and Brown, also one by Dennis Martin who was a member of the first wagon train into California.

On Los Gatos Creek, which flowed out of the east flank of the Santa Cruz redwoods, "Buffalo" Jones made lumber in a small sawmill. To the north of San Francisco Bay at Bodega, a steam-powered sawmill was established in 1843 by Capt. Stephen Smith who brought the machinery from the East Coast.

And so the Pattersons and the Smiths and the Hinckleys had their brief but stout-hearted flings at thinning out the redwoods but the big trees never felt the feeble hacking at their trunks. They would stand as a bulwark against the land and as sighting beacons for ships until Jim Marshall kicked up a nugget at Coloma in 1848 which unsettled a nation and settled a state.

SEQUOIAS OF SONOMA

(In *Wood and Iron,* 1895)

Much has been said and written concerning the giant redwood trees of California. The Sequoia groves of Mariposa, Calaveras and Tulare counties are famous the world over as the colossals of the vegetable kingdom; but it is not generally known that the tallest standing tree yet discovered in America was the product of Sonoma county. This tree grew upon the west bank of Fife Creek, just opposite the town of Guerneville, and was known to all of the early settlers on Russian River as "The Monarch of the Forest." It was one of the finest specimens of red-wood that has ever been seen by man. It measured 45 feet in circumference at the base and was 367 feet and 8 inches tall. This tree was felled about twenty years ago by Heald & Guerne and converted into lumber at their mill in Guerneville.

There was another very large tree about a mile north of Guerneville. This tree was sawed into lumber in '75 or '76 by the late Rufus Murphy, and yielded 78,000 feet of lumber, of which 57,000 feet was clear. The market value of that tree when converted into lumber was $1830, besides thirty or forty cords of wood, which at that time was valueless. The Baptist Church of this city was built wholly from the product of that one tree.

On the south bank of Russian River, on the land of the late S. H. Torrence, formerly stood the dead stub of a once mighty tree. This stub was broken off about two hundred feet from the ground, and was ten feet in diameter at the break, and was hollowed out like a huge flue or chimney. The writer and another party felled this stub about twelve years ago, and manufactured from it over $900 worth of stakes and other products.

THEY CUT 'EM HIGH to lay 'em low at Hoak's Ranch, Comptche. It was a week's work to lay the bedding, build a scaffold and fall a redwood in 1860. (Photo from Escola Collection)

THE HORSE WENT 'ROUND AND 'ROUND to yard logs by this crude winch at San Vincente Lumber Co. in the Santa Cruz country. (Photo University of California—Agriculture Ext. Collection)

But the largest tree that ever grew in Sonoma county, so far as is known, formerly stood on the bank of Russian River about a mile above Russian River station. This tree was twenty-three feet in diameter at the base and stood over three hundred feet high. It was chopped down by a man by the name of English, and was manufactured into shingles, of which it made upwards of 600,000, and afforded him labor for more than two years. The shingles when sold brought their maker $1800. This tree was so large that English was unable to saw the log in proper lengths, and was compelled to saw out a cut at a distance of two hundred feet from the base, and where the log was still over twelve feet in diameter, and with a maul and wedges he split the huge log (over two hundred feet in length) in halves. Had this tree been sawed into lumber it would easily have produced 100,000 feet, which would have been worth over $2000.

But these trees are not alone interesting on account of their immense size or the market value of their products. Surrounding them is much food for reflection and thought. These trees, like the rest of the products of the vegetable kingdom, "grow," or, to speak more accurately, "increase" in size by an annual deposit being added to the outer part or exterior surface of the tree. This deposit of woody fibre is very rapid during the spring and summer, but gradually becomes slower during the fall until winter, when it practically ceases. This process gives the end of a log sawed from one of the trees the appearance of being formed of alternate rings in circular layers of soft and hard fiber of wood, and one ring of soft and one of hard wood when taken together show the growth of the tree for that year, the soft wood being the deposit during the rapid growth of spring and summer, and the hard wood representing the slower growth of fall and winter. In the young trees this deposit is very rapid, often being more than an inch in diameter in a single year. This is especially true with second-growth timber, as may be seen from the sprouts which grow up around the stumps of trees that were felled by the Russians near Fort Ross in 1812. These trees show a prodigious growth, many of them having already attained a diameter at the base of three and one-half feet. But with the original stock the growth has been much slower, and as the tree grows older its growth is much less rapid, many of the larger trees scarcely gaining an inch in diameter in twenty-five years.

From what has been said, the reader will observe that these large trees are of very great antiquity. The largest tree that I have just mentioned was shown to be over 3300 years old. This tree was brought into existence while Moses and the Israelites were wandering in the wilderness, and was a thousand years old at the time of the birth of Alexander the Great. It was over seventeen feet in diameter (as shown by the annual growths) when Christ was upon the earth.

But ancient as the grand old trees are, they are but the modern growth that has sprung from the ruins of a larger and more ancient forest. It will be observed upon examination of these trees that they usually grow in circular or semi-circular groups, with a vacant space in the center of from fifteen to thirty feet in diameter. This vacant space was formerly filled by a pre-historic tree, and the trees forming the group are but shoots from the roots of the decayed monster. The writer remembers that about twenty-five years ago parties were boring a well in the midst of the forest which then covered the present site of Guerneville, and at the depth of thirty-eight feet they struck a redwood log and bored through it a distance of six feet. The tree had grown to its maturity and had fallen, and over its prostrate trunk the accumulated sands of ages had been cast until thirty-eight feet of earth covered its giant form, and then, 3000 years ago, the huge forest that so recently covered the place had its birth upon the grave of the ancient.

Another peculiar thing about these trees is that you often find acorns imbedded in the solid wood of a tree at a depth of four or five feet from the circumference. I remember distinctly of seeing a log sawed at the mill of Guerne & Murphy several years ago, in which a large number of acorns were found imbedded in the solid wood at a depth of five feet from the circumference. The acorns were deposited there by the pre-historic woodpecker, and although fifteen hundred years have elapsed since the little bird laid up his store of food for the ensuing winter, the descendants of his species are still there and are possessed of the same habits and instincts that characterized their ancient predecesors.

But one of the most remarkable instances resulting from the habits of these little birds of storing seeds of various kinds for preservation in the redwood tree, was observed by the writer when a boy. As late as 1870 there stood on the land of S. H. Torrence, near the Great Eastern mine, the black and charred stump of an immense tree. This stub was about two hundred feet high and six feet in diameter at the top. The top of this stub was evidently hollow at one time, and in the hollow had fallen the leaves of the surrounding trees until it was filled, and when in the course of time there leaves decayed and left the stub filled with rich and productive soil, some little bird had dropped a manzanita berry in the soil thus formed, and the seed took root and grew until the bush covered the entire top of the stub with its branches, and every year this bush was filled with beautiful snow-white blossoms, and resembled an immense bouquet set in a large bouquet.

FRAGILE FOREST GIANT — This Humboldt County redwood fell during a storm and was shattered on the ground. Professional loggers laid bed of branches to cushion shock. (Ericson photo from California Redwood Assn.)

GOLD UNDER THE RED BARK

Gold had the color and the glory. It had the fascination and the power to pull people into California from every state and territory. The big yellow chunks and the little yellow grains spelled romance, sparked the imagination and promised hell-roaring adventure. The promise was fulfilled — but pokes went empty.

Gold was found but it was also found wanting. If it made one man wealthy, it broke ten thousand. In the ten short years following its discovery, it uprooted and reshaped families, built and ruined cities, set a new pattern of life for the new State of California. The suddenly swelled population created a heavy burden but here in this new land there was opportunity for everybody who wanted to work and for the more enterprising who had a sense of business.

A hundred men in the transplanted body had both enterprise and the will to work. They saw in the redwoods a big challenge to their ingenuity. Here was stability where gold was quicksand. Here was permanence, a livelihood for many, a fortune for a few, both of which should continue through the years. They said:

"Look at all those big, beautiful trees. You don't have to go hunting for them. You know the

NEWPORT CHUTE IN THE '70s (opposite) at Stewart, Hunter and Johnson's mill near these Pacific cliffs. Schooners anchored by two or more lines with enough slack to allow running twenty or twenty-five feet with the sea. At the lower end of chute was "clapperman" who operated a brake-like device to slow up and stop each stick. Schooner of this size could load 75,000 to 150,000 feet. (Photo Union Lumber Company Collection.)

TIE WHACKS MADE A DAY'S PAY — The wood cost nothing, arms were strong, time endless. Ties were carved out of the smaller redwoods near mills on the coast and hauled to schooner loading points by horse team. These men were working near Fort Bragg. (Photo Union Lumber Company Collection)

TIES MADE WHILE YOU WATCH
— Four tie whacks show full process. One man sawing logs to 8-foot length, man at left marking tie size on open cut, man with sledge driving wedges to split to approximate size and fourth shaping tie with broad-axe. (Photo Union Lumber Company Collection.)

supply won't run out. And with all these people needing houses and all these new cities needing building material, there will be no question of market demand. We'll have to learn how to cut and handle these big trees and we may need luck, but we can make lumber."

And so those strong-hearted men from Maine, Michigan, Pennsylvania, Minnesota, Ohio and elsewhere went to work, some of them with experience, some with capital. San Antonio, on San Francisco Bay, became a busy lumber port, the smoke from the steam sawmills shrouding the skyline and drifting down the canyons where little roads went from mill to market in Martinez, Alvarado and Sacramento.

William Taylor and James Owen had a mill in the Moraga woods where later there was the operation of Henry Thorn and William Hamilton. The mill of Thomas and William Prince was set in the canyon of Redwood Creek and that of partners Joseph Wetherall and Nathaniel Lampson was a little farther to the north. Those two went in debt but sold to Henry Speier who cut lumber until the Sonoma redwoods were exhausted. At the mouth of Redwood Canyon, Chester Tupper and Rich Hamilton built a mill. It burned in 1854 and was purchased by Thomas Eager and Erasmus Brown, a later partner being Daniel Plummer.

In the San Mateo region and the Los Gatos country disillusioned gold miners and adventurers from the East Coast were putting logging and sawmilling tools to work. Northward up the redwood coast men saw possibilities in the big timber. In 1850 Eddy and White operated the Taupoos Mill for a short time. Then the beached steamer *Santa Clara*, was converted into sawmill power. Timber was rafted in, the mill hands slept aboard ship and the mill cut thirty thousand feet of redwood a shift. This was the start in Humboldt Bay and by 1854 there were nine sawmills here vying for trade.

The backwash of the gold rush produced in San Francisco another figure who deviously and briefly was to give impetus to the advance on the redwoods. This was Harry Meiggs, of colorful and magnetic personality, who at thirty-eight in 1849 had a reputation for honesty, was ambitious and the kind of fellow the city's bankers liked.

On a day in 1850, Harry Meiggs saw stacks of redwood lumber on a wharf and learned it had come down from the Mendocino Coast. He got reports on the standing timber there and was at once intrigued by the vast possibilities in harvesting it.

He was not alone in this fascination. He found E. C. Williams, Jerome Ford and Capt. David Lansing all interested. The first of this trio was an alert young New Yorker, who after the war with Mexico and his own personal war with California gold, found himself owning a lumber yard in San Francisco at North Beach, at the head of what was later to be known as "Meiggs Wharf." Edward Williams was ready to act and in the following account, which appeared in the Mendocino Beacon in 1912, gives his personal story.

WATERLINE TO RHONERVILLE sending ties three miles to schooner drop at Point Arena. (Photo Union Lumber Company Collection.)

"At the suggestion of Mr. Meiggs, I began an investigation which led to the discovery of a portable circular sawmill at Bodega on the ranch of Capt. Smith. This had a capacity of eight or ten thousand feet of lumber in one twelve-hour day. Mr. Meiggs bought the plant but soon learned that its limited capacity did not accord with his broad ideas and he dispatched me East (the Isthmus route then being open) to have built an up-to-date gang mill with steam engines and boilers, to be sent by sailing vessel via Cape Horn.

"I found that I could get such a mill as I wanted at Painted Post, a small town in the State of New York, near the southern boundary, but the eastern states would give me the engine and boiler cheaper and quicker than they could be built elsewhere, consequently, these were bought in Norwich, Connecticut, several hundred miles distant from the shops building my mill machinery. In studying my plans I found that it was necessary to make some changes in the figures connecting the motive power with the mill machinery. As a very unusual occurrence, Long Island Sound was frozen over that winter, so for some time the steamers could not run, and the railroads throughout the north and east were very irregular and retarded in their operations. To do what I wished it was necessary to hold immediate communication with

Norwich and Painted Post. In stating my dilemma to my brother, he said, 'Why don't you telegraph?' 'Telegraph,' I replied, 'how do you do it?'

"'Why,' he said, 'send a message to each of the shops to stop all work on the connecting parts until they receive your letters of this date.'

"This I did and the trouble was avoided. Of course, I knew there was such an instrument as the telegraph, but having lived for years on this coast, where travel went on foot or horseback, it did not occur to me to employ the new methods which had come into use since 1846 when I had left New York.

"Having finished my business in the east and seen my purchases on board ship, I returned via Panama and reached San Francisco to learn that investigation by Mr. Meiggs had convinced him that such a mill as was on the way from New York would soon exhaust the supply of standing timber at Bodega, and he had taken steps to find a better location. Inquiries in this direction brought him into contact with William Kasten, a German gentleman who had settled at the mouth of Big River, and was enthusiastic in his praise of the timber on its banks and of the opportunity of milling and shipping. On his report Mr. Meiggs had chartered a small schooner and sent her up the coast to report on the situation, and was expecting

25

TOOLS OF THE TIE WHACKS — Thirty dollars put you in business cutting redwood ties but only endurance kept you in. (Photo Union Lumber Company Collection)

the party back every day, but the day after my arrival a messenger brought a letter from one of their number, Mr. J. B. Ford, stating that the schooner had been unable to make headway against adverse seas and winds and had come back to Bodega Bay to wait for more favorable weather. I started back with the man who had brought the letter and joined the party on the schooner."

(Meanwhile J. B. Ford had decided to go overland to meet Williams. From Capt. Steven Smith at Bodega he bought eight oxen and hired two men. The party worked its difficult way up country on rough, narrow trails, swimming the swollen rivers. Crossing the Gualala, one mule was drowned and another swam away, the party losing all its provisions. Sleeping under "sweat cloths," as the saddle blankets were called, going without food for thirty-six hours, it reached the 'Portuguese Ranch' near the Navarro River, replenished stomachs and saddle bags, arrived ten days later at Big River.)

"For several days it blew a gale from the northwest and I took the opportunity to visit the mill at Bodega and saw at once the place was wholly unsuited for the new enterprise. In a few days the weather favored us and we reached the haven of our desires. Mr. Kasten had gone overland.

"In a rough canoe, which he had fashioned from a redwood log, Mr. Ford and myself started out on a bright April morning with a fair light breeze from the ocean and a flood tide to prospect the river for timber. For the first half-hour we were rather disappointed, but after that all that we had hoped for was more than realized.

"The winter rains had not wholly ceased and the river bank full, its slight ripples meeting the verdure of the shore, the tall redwoods with their great symetrical trunks traveling toward the skies; with the bright colors of the rhododendrons profusely scattered over the hills forming the background, the clear blue sky above reflected in the placid river and over all the hush and solitude of the primeval forest—all combining to impress upon our minds the beauty and truth of the opening of Bryant's Thanatopsis, 'The groves were God's first temples,' and as I recall the beauty of the picture, I cannot but regret the part it appeared necessary for me to enact in what now looks like a desecration.

"Our dugout was a heavy awkward thing to handle and I think we had gone up the river not more than three and a half miles, when on the first ebb tide we began to paddle homeward. This was a very different thing than our earlier experience. With wind and tide setting forth our way, we had but little difficulty in going up, but on reversing our course we had a strong tide running out, helped by the accumulated fresh water which the flood tide had kept back, and, as we approached the coast, the wind which was blowing up the river created a swell which threatened to fill and swamp our canoe with its scant two inches of freeboard. About one and a half miles from the ocean (as we judged) we tied up the canoe to the shore and started straight up the hill on the north side of the stream. After some scratched faces and bruised knees and shins we reached the top of the ridge and soon found a well-worn Indian trail which led westward, and before very long, saw the ocean again, much to our joy. We had seen enough to satisfy us and the next morning with a fair wind, sailed for San Francisco, where we arrived without further misadventure.

"Upon the enthusiastic report of Mr. Ford and myself, steps were at once taken to prepare for the arrival of the machinery and its transportation up the coast. There were at this time many vessels in the harbor which had been deserted by their crews who had gone to the mines some months previous. One of these — the 'Ontario' — a ship of about 500 or 600 tons, was bought, and on the

LITTLE RIVER, MENDOCINO COUNTY (*above*) View from log landing. House at right is first built here — 1857. Coit home at right center on river. Above this is store and white building above store is Mahlmann Hotel. (*below*) View of mill owned by Coombs and Stickney, 1863-73. (Photos Bancroft Library, University of California)

arrival of the machinery it was transferred to that vessel. With a force of about forty men, including mechanics and laborers engaged, early in June, 1852 we set sail for Mendocino, the ship being in command of Capt. David Lansing.

"We were not out of sight of land before our troubles began. The ship, which had been at anchor for a long time, had become very dry above the waterline, and as soon as put into the wind, and headed up the coast the motion began to loosen the oakum in the seams and let

CAMP AT FORKS OF LITTLE RIVER and mill (*below*). Top and bottom head saws with crew — at left of saws Phil Doust, sawyer; behind him, John Norberry, setter; behind him, John Dennen. At right, Pete ——, offbearer; Peter Anderson, Chris Anderson (with whiskers), Bill Blair (with mustache). (Photos Bancroft Library, University of Calif).

in the water, not very badly at first but increasing from day to day until the workmen became alarmed and insisted upon going back to San Francisco. The Captain said that with good weather there was no danger, and the workmen agreed that if they were put on pay they would man the pumps and keep the ship free which was arranged for, and the wind favoring, we made port without mishap.

"We found Mr. Ford, who had started some days in advance of our sailing with horses and oxen, already on the ground, and as the summer winds had begun we hauled the ship inside the point with the stern close to the shore and landed her cargo. Later the hold was filled with rock from the bank until the hull was fast on the bottom, our intention being to complete the filling of the hold and 'tween decks later and thus make a permanent wharf and breakwater. This however, was deferred too long and the first storm separated the upper works from the heavily-weighted portion and washed it up on the beach. Ultimately, the mighty wave force washed the rock from the hold and some of the old timbers were to be seen long afterwards.

"The difficulties with the building of the mill were many and great. Our millwright proved wholly incompetent, men became dissatisfied and left at a moment's warning, and their places could only be supplied by our sending to San Francisco and bringing them overland up the coast. Before the mill had its roof on, the storms began, and the memory of that winter for years came to me as a horrible nightmare; but spring came and at last the mill was finished, and we were shipping its output to market at the rate of about 50,000 feet per day. We found that a gang mill was not suited to the size of our logs or the character of the timber; nor did it give Mr. Meiggs the quantity of lumber he wanted ,and he told me to give him 100,000 feet a day as soon as it could be done. Accordingly, a new plant was built on the site of the present mill, turning out 60,000 feet in twelve hours, operating circulars instead of a gang. The engine for this mill and all the machinery were made in San Francisco. It was destroyed by fire a few years later and rebuilt on the same general plan.

"To go back a little. The cost of building in excess of the calculation, coupled with the delay of getting any returns from the lumber sold, that in 1853 The Redwood Manufacturing Co. had been incorporated with a capital of $300,000 the stock of which Mr. Meiggs had used as security for his loan to a large extent. A few months after the new mill was built business in San Francisco became very dull, and Meiggs became involved in financial troubles and left the country. In order to carry on his extensive operations he was a borrower of large sums and for this purpose he had sold some 2,000,000 feet of lumber of W. N. Thompson to be delivered at a later date and had received partial payment in advance. The company also owed Williams and Ford, who were operating the mills under contract to deliver their product on board the vessels at a stipulated price. He also had borrowed heavily from Godeffroy, Silem & Company, then prom-

SCHOONERS LOADING FROM APRON CHUTES at Mendocino Harbor in 1865. (Photo Union Lumber Company Collection.)

FIRST WHARF AT LITTLE RIVER — 1860s. (Photo Bancroft Library, University of California)

inent bankers in San Francisco giving the stock of the company as collateral. These conditions of course involved the company in bankruptcy and the mills were idle for a long time.

"The final result was that the bankers and Williams and Ford were given possession of the property, upon condition of giving W. N. Thompson the lumber his contract called for, and was organized in 1872 which was continued until it was dissolved by the order of the court referred to at the beginning of this article. It may be of interest to lumbermen and to forestry experts to know that from the 36,000 acres embraced in this holding the company had cut 7,150,000,000 feet of lumber, and from the average estimate of several cruisers there was still standing at the date of our selling 450,000,000 (I think it will prove to be 500,000,000) feet, which would give 42,000 feet per acre. This I think is a fair average of large bodies of redwood timber in Mendocino county, though there are some quarter-sections that will yield double that quantity.

"The history of the first shipment of redwood to a foreign market may be interesting to some of the men who have later come into the business and it is here given for their benefit.

TWELVE FOOT SAW FOR TWELVE DAY JOB (*opposite*) This big Humboldt County redwood has been felled and now comes the staggering job of bucking it to shingle bolt length. Two complete cuts were good day's work for two husky men. (Ericson photo from California Redwood Association)

"In 1857 or '58 the Mendocino mill shipped to Sydney 100,000 feet, mostly surfaced and T&G clear boards, consigned to correspondents of Godeffroy Brothers of Hamburg. They paid freight and all subsequent charges of storage, fire insurance, etc., and some ten or twelve years later sent the mill a statement closing the account showing a debit balance of something less than $50 against the shipment, which they under the circumstances, generously remitted.

"The trouble with the lumber arose in the first place from the color, for the people there were very conservative; and in the next instance from the fact that all their nails came from England, and were pointed from the head down on all sides. Of course, these nails — no matter how they were driven relative to the grain of the wood — acted as a wedge and split the boards, which confirmed the condemnation of this 'blarsted stuff.' But, as the years went on, and the market was at times depleted of other woods, and lumber of some kind must be had, small lots were disposed of and some of the good qualities began to be known and appreciated. And still later, when cut nails from the United States came into use, redwood increased in favor, and our loss of $2500 opened the door for the larger and more profitable market now enjoyed by the shippers to that country.

"In closing these reminiscenses the writer is glad to think that the hardships and trials of the pioneers in the business are passed. That chapter in the redwood industry is closed — and so are these reminiscenses."

31

SCHOONER LOADS AT MENDOCINO — 1852. Chutes send lumber down to steam schooner at point out from Big River. First mill of Mendocino Lumber Company at right, out of picture. (Photo from Vaughan-Escola Collection)

DEL NORTE---TO THE NORTH

The coastline up from the booming Golden Gate country was settling fast. Schooners brought people and money for exploitation, continued on up to Astoria and Puget Sound for cargoes of fir, returning for what redwood they could get.

"Answering the first local demand for lumber," writes Leona Hammond in the Del Norte Triplicate Centennial Edition of 1954, "F. E. Weston built in 1853 the first sawmill in Del Norte county, locating it in the still-present gulch near the junction of Third and C streets.

All redwood logs used at this first mill were hauled from what is now known as Howland Hill. They were reported to have been hauled "on two large wheels about twelve feet in diameter." After running a little over a year, the mill was moved to G street opposite the one-time residence of the Hon. W. A. Hamilton. When the mill burned in 1856, a Mr. Kingsland removed the usable machinery to his new mill near Elk Creek.

"The next mill to be built was that owned by W. Bayse, which was located on Mill Creek, about six miles from Crescent City. It was operated by water power. A road was constructed over the hill for the purpose of hauling lumber into town, but the cost of transportation was so great that after a few years the enterprise was abandoned.

"Also recorded in the early 'ups and downs' of the sawmill industry in the county was the horse-power mill which was constructed near where the Elk River mill later stood. But, according to an unidentified newspaper clipping of the 1890's, 'as the carpenter who was building the City Hotel was able to pack the lumber from the mill to

the hotel and work it up as fast as it was sawed, it was not considered a good investment and was abandoned.'

"A dropping off in the freight business conducted by steamer and sailing vessel to Crecent City was noted in 1870, with the advent of wagon roads. Then, too, came the demise of all the pack mules. Until that time, all goods needed in northern California and southern Oregon had been shipped to Crescent City. Supplies for the interior were carried on pack mules, and it was not uncommon to see 150 mules packed in one day.

"As a result of a public meeting called by farsighted individuals, the first cooperative sawmill effort was instigated as a means of promoting waterfront trade activity, and at the same time, utilizing the immense stand of redwood and spruce so providently accessible.

"Although a wharf to facilitate lumber shipments was soon built by Justus Wells and J. K. Johnson, the first shipments were made in a cumbersome, but unique way. The lumber was hauled in wagons from the mill on Lake Earl, to the waterfront, where it was piled above high water mark; a sufficient number of rollers were made to reach about 200 yards, placed three feet apart, which ran the lumber to the lighters.

"The first lumber cargo carried from this port by the schooner Fanny Jane, with Peter Caughell as master, was the forerunner of Crescent City's present-day lumber shipping industry. After several cargoes had been sent to San Francisco in the manner described above, the lighters were loaded alongside the wharf.

"The mill operated with success until sometime in the 1890's when it was destroyed by fire, proving a severe loss to the community.

"In 1866, the attention of outside lumbermen was first attracted to the region. A mill and box factory was established on Elk Creek by Caleb Hobbs and David Pom-

Second saw mill -- 1854

SECOND MILL AT MENDOCINO. With demand of 100 thousand feet of lumber a day, the Mendocino Lumber Company built a second mill of the river flat in 1855. A few years later the mill burned (*below*) but was immediately rebuilt. Both photos were taken from same camera position by Carleton Emmons Watkins. (Bancroft Library, University of California.)

INCLINE TO SHIPPING POINT. Mendocino mill on river flat sent lumber up this tramway to chutes, thence to steamers. (Watkins photo from Bancroft Library, University of California)

eroy of Hobbs, Gilmore and Co. J. G. Wall became associated with them, and the firm became known as Hobbs, Wall & Co.

"The mill operated with apparent success, the company building a railroad to Smith River, and operating two steamers, the Crescent City and Del Norte, for the lumber shipping trade. Its steamers also carried passengers and freight. The company ceased operations here in the immediate pre-war period.

"Roy Ward, retired former worker for Hobbs-Wall, who was on the payroll at the time of its demise, last man on the lumber company, has vivid recollections of the logging railroad which extended from the Elk Valley mill in Crescent City to the Howland Hill district.

"He recalls the jealousy between two identified locomotive firemen, which finally erupted in a lively incident in which he, as engineer, was an unsuspecting participant.

"As he and one of the warring firemen approached the old trestle, which still stands adjacent to the Humboldt road behind the Bertch tract, the other fireman lay in wait for the slow-moving logging train. When it reached the spot where he was concealed, the man flung a hornet's nest into the cab.

"The unlucky fireman really 'fired up' the boiler at that point, and Ward and he, both severely stung, made a record run back to town, much to the amusement of the man back in the brush.

"Mrs. Ward's recollections of that period are of a somewhat less humorous nature, although they, too, bear witness of the import of the Hobbs-Wall operation in the community.

"Mrs. Ward was one of the local residents accustomed to making occasional trips to San Francisco aboard the lumber-carrying steamers owned by the company. These trips were quite frequently made during stormy weather, and at such times returning passengers, she recalls, were unceremoniously hoisted ashore in a barrel slung from a line extending to the dock from the ship, which would be anchored in safety some distance away.

"On one such trip, she says, a body was being returned to Crescent City for burial, and the passengers had been uncomfortably aware of its proximity during the entire voyage up the coast from San Francisco.

"Upon arrival, conditions were stormy, and the steamer anchored out from the dock. When the line was rigged, she was startled to find that the first 'passenger' ashore was to be the one whose remains lay in the casket. She vividly recollects her feeling of fascinated horror as the barrel in which she was later to ride to shore slowly made its precarious way along the line, carrying its gruesome cargo.

"It reached the dock safely, she says, but her own trip in the barrel had lost its novelty.

"Several other mills were in existence in this natal period. A. M. Smith operated a mill on Smith River near the former railroad crossing, and the Fairbanks brothers in 1859 built a mill near Smith River Corners, in which N. O. Armington was aso interested. J. G. Anthony operated an adjoining flour mill.

"At one time, Henry Westbrook, John Bomhoff and R. D. Hume operated a mill near the mouth of Smith River, with a railroad and logging camp in connection with it. Victor Ohlsen owned a mill operation at Fort Dick;

FALLING TWELVE FOOT REDWOOD Note area cleared behind tree with bedding laid to cushion fall. (Photo Union Lumber Company Collection.)

CASPAR CREEK MILL IN 1875. Built by Kelly and Randall in 1861, it burned in 1889 and was rebuilt. (Photo Escola Collection)

Bertsch Bros. had a shake mill near Smith River and Bailey Bros. owned a similar mill near the Smith River railroad crossing.

"During 1875-76, Thomas Van Pelt operated a small mill back of Pebble Beach, hauling lumber to Crescent City to be shipped to San Francisco.

"On the Klamath in 1890, Schnaubelt Bros. operated a local mill. In the same year, one sawmill and two shake mills were operating in Crescent City, and the newspaper of the day expressed the hope that 'in the near future, there will be at least a dozen of each, as we have an almost inexhaustible supply of good timber in the country.'"

LANDING FOR BURKE'S TIES (*below*) at right of Mendocino mill. Burke had tie camp up river above company ranch. Blanks were floated down to foot of Big Hill, loaded on cars and hauled to boom, thence to lighters and mill. Horses then hauled finished ties up apron where they were winched up incline, then hauled by horses to point and sent down chutes to steamers. (Photo from Escola Collection)

CALIFORNIA SAWMILLS: 1853

The following anonymous account appeared originally in the Northwestern Lumberman and was reprinted in San Francisco's Wood and Iron.

In January, 1853, I arrived in San Francisco, Cal., with just $3 in my pocket. I had to have a job at something for the cheapest board was $10 a week, and there were more than 5000 idle men there, so I went among the re-sawing mills and at last found one where I thought they needed a sawyer. It belonged to two old sea captains, Bailey & Hammond. Bailey was the manager, and I noticed at once that his sawyer knew nothing of the business and was doing slow and very poor work. There were two small circular saws, and when one was at work the other was idle, as they had not power enough to run both. The timber came mostly from Oregon, and was heavy plank 34 and 36 inches thick, and we had spruce that was woolly and had black knots about as hard as a spike, and a species of yellow pine.

I applied to Captain Bailey for a job and the gruff old fellow said: "No, sir; we've been humbugged all we are going to be," so I walked off. But in passing the little office I saw Captain Hammond and asked him if he was one of the proprietors, and was told that he was. I related my experience, and, Yankee like, bragged of what I could do, and finally said: "Can I get your permission to come in, put one of these saws in order, and make no charge whatever, not even for files, and let me work the balance of this week, about three and a half days, for nothing?"

His reply was, "Oh, yes." So off I went and got a pair of overalls, and went into a blacksmith shop and made me a small upset or swage, and paid $1 for a ten-inch file.

Next morning I was on hand and walked right up to the idle saw and went at it. Pretty soon Bailey came in and gruffly said: "Who the d—l set you to work here?

We don't want you." I began to tell him, when a purchaser called, and away he went to sell some lumber, and one customer after another kept him for an hour or two. I went on and put my saw in order, and seeing a pile of spruce lying there which they had tried to saw but failed, as I afterwards learned, I slipped around to the engineer and asked him what they wanted it sawed into. He told me half-inch siding. So as soon as their other saw got dull and they threw off their belt, on went mine, and I began sawing. I had no trouble in sawing it true and all right. Engineer Ford called Bailey's attention to it, and I soon saw the old fellow peeking around the corner of the little office and grinning. He did not come near me, but had the engineer tell me what to saw. The next thing the other sawyer was out, and two big darkies were handling and piling lumber.

A few rods from there was another re-sawing mill, and the proprietor came up to me one day and offered me $10 a day to run his mill. I told him I would let him know Saturday night. So on Saturday night I gathered up my tools and started for my boarding house with no money to pay my board, which was due every Saturday, but our engineer offered to lend me what I wanted, so I was independent.

After I had got a block or so away I heard someone calling: "Halloo, there, halloo!" I turned around, and the old short captain came up puffing, and said: "We pay every Saturday night." I said: "I have no bill against you; I agreed with your partner to work this week for nothing." "Oh, well," said he, "now I will own up that we never had any sawing done until you came."

I walked back, and he said: "How much do you want?" Said I: "Well, I am offered by your neighbor $10 a day, but did not agree to accept it." He threw down four $10 gold pieces, and said: "That pays you for three days, and the other $10 is for putting the saw in order." I worked for them for eight months every day, and often two or three days in a week overtime.

DUMPING LOG THE HARD WAY— by jackscrew in 1875. Tom Dollard and Arthur Jarvis have pushed truck down pole road to river at Jim Nichol's camp at Big Hill, Mendocino County. Dollard was later killed by outlaws. (Photo Escola Collection.)

The stories of fabulous wealth in the gold mines, the piles of gold on the gambling tables, and my pockets filled with $50 slugs, made me uneasy to go to the diggings, as they were called. So I quit work and went to Grass Valley, then one of the famous mining districts, and after paying $1 for a meal and lodging, and poor at that, and looking around, I was not venturesome enough to go into business. But as lumber was wanted for mining and standing timber was free, I looked around and found a mulay saw mill, belonging to a gentleman named John C. Birdseye. I took it to run by the thousand, and he supplied it with logs. I added a small circular for making siding.

After getting well under way and making some money, it took fire, and all went in the night. Next day I bargained for the site, the boilers and the engine, which were not badly hurt. I started for San Francisco and went to my former boarding mistress, and borrowed $2000 from her, took in a partner named Eddows, bought a very poor circular saw mill rig, the best I could find, and other things needed, such as belting, etc., and in about six weeks had a mill that would turn out four or five times the lumber of the old mulay. The former owner's brother hauled the logs with a yoke of oxen that he called Tom and Jerry. The old man liked his brandy, but invented a novel and rather successful way of not over drinking. He got a quart bottle of brandy, and every time he took a drink he filled it up with water, so that it grew weaker and weaker. After several months I said: "Birdseye, I can't taste any brandy in that." "Well, no, there ain't much brandy taste there, but it tastes a little better when it comes out of that bottle."

There were soon about thirty mills around Grass Valley, so that lumber became cheap, and there was so much of it that we opened yards at Marysville and Sacramento. Marysville was thirty miles and Sacramento sixty miles distant, but as it all went on return teams that were hauling supplies up, freight was considered quite low, $18 to $25 a thousand. Owing to the cost of keeping up a yard and freighting, our price at the mills netted us but a small sum, so that many mills went under. In time, however, as timber became scarcer, and farther to haul, lumber advanced. Farmers fenced off farms, so that soon we had to buy our standing timber. I took time by the forelock, and secured about three million feet of standing timber, on which I cleared more than $5000.

I had saved $3000, and took a Wells, Fargo & Co. draft on San Francisco, in order to pay Mrs. Watts what I owed her and to make some purchases. I was on the outside of the stage going to Sacramento, when we saw a man coming on horseback at a dead run. As he came up to us he called out, "Have you any express matter?" The driver said, "No; it is on the stage coming." He then said, "All of the banks have closed, Adams & Co., Wells, Fargo & Co., and all," But I felt quite secure, for I believed that Wells, Fargo & Co. were all right. I told Mrs. Watts so, and she did not seem to worry much about the money.

I sent a note to the agent of the company by one of their clerks, with whom I boarded, and the next day he said that they had sent gold dust enough to the mint to be coined to pay every depositor and draft, and that if at ten o'clock the next morning I would go into a certain

narrow alleyway and to a certain window at the back of the building, and rap on it, they would pay my draft. So I went, and got the amount in gold coin. In a few days the bank was opened and they paid everything off. Adams & Co., as bankers, never paid anything.

One gentleman of my acquaintance had a singular experience at that time. He was a blacksmith, and had an order for a lot of bolts that required washers of quite a large size. He went to an iron firm, Conroy & O'Connor, and got a lot, and put them in a bag, and called at Adams & Co. for some letters which came by express. When he came out a gentleman said, "I see you got yours all right." "Yes," said the blacksmith. "Well," said the man, "so did I, and I don't know where to keep it." He

SEMPERVIRENS CLUB posts burned out center of big sequoia. Fire probably struck tree many centuries ago but tree continues to grow and remain in good health. (Photo Bancroft Library, University of California.)

followed him to his shop and put $7000 or $8000 in his safe, and finally lent it to him a year without interest. The blacksmith took it to pay for appealing a land case to the Supreme Court, which he won. He sold the land for $7,000. I saw him in San Francisco three years ago. He is now wealthy.

Timber became quite scarce in and around Grass Valley, so I sold the Penobscot mill and went on the divide between the Yuba and Feather Rivers and built a large circular saw mill in the sugar pine, which is a species of white pine, but of much larger size. Many of the trees were twelve to fourteen feet through at the stump, and so free rifted that I often bored a small auger hole in them and blew them open on the rollway. My head-blocks were about twelve feet long, so that I could saw the largest of them. I ran a 72-inch lower and 66-inch upper saw, and then many of the logs I notched down on the head blocks so as to get off the first slab.

I was then in a region of immense timber of fine quality, where there were bears and plenty of wildcats, and occasionally a cowardly California lion. Often in the night the wildcats kept us awake.

I had never visited Humboldt Bay, which was head-quarters for the redwood timber, and no one then ever imagined the vast territory of more than 100 square miles of timber that would average over 300,000 feet to the acre, for no human being nor animal of any size above a rabbit could get through the thick bushes. Three years ago I visited the Pacific Coast and went to Humboldt Bay. We have furnished nearly all the mill owners there with saws for years. Major Vance of Eureka, who owned three mills, took me out on his logging engine about eight or ten miles to the forests, and there he had three engines, a team of three yoke of oxen and another of six mules loading logs to go to the mill. As I got there he was loading one monster, and I asked him how much it would weigh. He said probably about fifteen tons.

The butts of nearly all redwoods are splintered at from six to ten feet, so they mortise in and put in a small staging. On this the men stand and saw the trees off with crosscuts. The bark on them is from 12 to 24 inches in thickness. The method of working is to bush through in winter, saw down the trees, and when bark peels in the spring peel them as far as the timber is good. Then in the dry season they set fire to the dry underbrush, and as redwood is hard to burn it is not injured. The fact is, it is a heroic job to get this timber.

HUMBOLDT COUNTY REDWOOD was fallen over to down log to raise butt off ground. Choppers have earned rest after four or five days' work. (Photo Hammond-California Lumber Company Collection.)

THE BARRIER BEGINS TO YIELD

You took the steamer of the North Pacific Coast Railroad out of San Francisco and landed on the north side of the Bay where soon there would be a town called Sausalito. The narrow-gauge rails ran over the sandy hills around the base of old Tamalpais with San Quentin and San Rafael to the right. And at Camp Taylor you got your first view of the big redwood trees. You cleared the cattle country of Marin, climbed the mountains and descended to the valley of the Russian River. You made this pilgrimage if you wanted to work at Duncan's Mills in the heart of Sonoma County's redwoods in 1880.

There had been some people ahead of you since Capt. Stephen Smith started his sawmill at Bodega in 1843. On the Jonive Rancho a water-power mill had been built by "Blinking Tom," Edward J. McIntosh, James Black, Thomas Butters, William Leighton, Frederick Hegel and Thomas Wood. In 1849 they sold the mill to F. G. Blume and went to the mines. Later that year a few carpenters and mechanics who had worked on the government barracks at Benecia found redwood selling at three hundred dollars a thousand feet, went into sawmilling — buying timber of Blume and setting up a mill on Ebabias Creek. This was the Blumedale Sawmill and Lumber Co. — Charles McDermott, president, and John Bailiff, secretary. By the time the company got sawing the price of lumber dropped so low the mill was closed.

Lt. George Stoneman, with Joshua Hendy and Samuel M. Duncan, purchased this property and ran the sawmill for two years, until Stoneman retired and the firm became Hendy and Duncan.

CHINA GRADE MILL of I. T. Bloom, dynamic redwood pioneer in Santa Clara Valley. (Mrs. Hare photo from Bancroft Library, University of California)

THE BOARDS WERE WOOLY AND WEEWAWED but they built the early mansions. Early steam mill at Guerneville, Sonoma County. (Cherry photo from California Redwood Association)

In 1852 the mill was moved successively to Yankee Jim's mining camp, to Michigan Bluffs and finally at Salt Point it became the first steam sawmill in Sonoma County north of the Russian River. In 1855, Joshua Hendy sold out to Alex Duncan and the firm became Duncan Bros. who in 1860 moved the mill back to the mouth of the Russian.

This was one of the big operations of the day and talked about on Montgomery Street and in the waterfront saloons. There were tales of how the Duncans cut 100 million feet of redwood and how they would have raised that if the flooding river had not carried all those logs out to sea. The winter of '62 was the worst — over 7 million feet lost. And a Sonoma historian writes a lucid account of the Duncan success.

"The woodsman chooses his tree, then proceeds to build a scaffold up beside it that will elevate him to such a height as he may decide upon cutting the stump. Many of the trees have been burned about the roots, or have grown ill-shaped near the ground, so that it is often necessary to build the scaffold from ten to twenty feet high. This scaffold, by the way, is an ingenious contrivance. Notches are cut at intervals around the tree at the proper height, deep enough for the end of a cross-piece to rest in securely. One end of the cross-piece is then inserted in

the notch, and the other is made fast to an upright post, out some distance from the tree. Loose boards are then laid upon these cross-pieces, and the scaffold is completed. The work of felling the tree then begins. If the tree is above four feet in diameter an ax is used with an extra long helve, when one man works alone, but the usual method is for two men to work together, one chopping "right-handed" and the other "left-handed." When the tree is once down it is carefully trimmed up as far as it will do for saw-logs.

"A cross-cut saw is now brought into requisition, which one man plies with ease in the largest of logs, and the tree is cut into the required lengths. The logs are then stripped of their bark, which process is accomplished sometimes by burning it off. Then the ox-team puts in an appearance. These teams usually consist of three or more yoke of oxen. The chain is divided into two parts near the end, and on the end of each part there is a nearly right-angled hook. One of these hooks is driven into either side of the log, near the end next the team, and then, with many a surge, a gee, and a haw, and an occasional (?) oath, the log is drawn out to the main trail to the landing-place. If on the road there should be any up hill, or otherwise rough ground, the trail is frequently wet, so that the logs may slip along the more easily. Once at the landing place, the hooks at the end of the chain are withdrawn, and the oxen move slowly back into the woods for another log.

"The train has just come up, and our log, a great eight-foot fellow, is carefully loaded on one of the cars. As we go along the track on this novel train on our road to the

LEARNED THEIR LOGGING AT SEA. Making no undercut these early Mendocino settlers had to drive eight wedges behind saw to keep tree's weight off cut. (Photo Escola Collection)

mill let us examine it a little. Beginning at the foundation, we will look at the track first. We find that the road-bed has been well graded, cuts made where necessary, fills made when practicable, and trestle work constructed where needed. On the ground are laid heavy cross-ties, and on them a six by six square timber. On this an iron bar, about half an inch thick and two and a half inches wide, is spiked the entire length of the track. The two rails are five feet and five inches apart, and the entire length of the tramway is five miles. Now we come to the cars wich run on this queerly-constructed track. They are made nearly square, but so arranged that by fastening them together with ropes a combination car of almost any length can be formed. And lastly, but by no means the least, we come to the peculiarly-contrived piece of machinery which they call a "dummy," which is the motor power on this railroad.

"This engine, boiler, tender and all, stands on four wheels, each about two and a half feet in diameter. They are connected together on each side by a shaft. On the axle of the front pair of wheels is placed a large cog-wheel. Into this a very small cog-wheel works, which is on a shaft, to which the power of the engine is applied. There is an engineer on either side of the boiler, and they

have a reverse lever, so that the "dummy" can go one way as well as another. By the cog-wheel combination great power is gained, but not so much can be said for its speed, though a maximum of five miles an hour can be obtained. On our way to the mill we passed through a little village of shanties and cottages, which proved to be the residences of the choppers and men engaged in the woods. Farther on we pass through a barren, deserted section, whence the trees have all been cut years ago, and naught but their blackened stumps stand now, grim vestiges of the pristine glory of the forest primeval. Now we pass around a grade, high, overhanging the river, and, with a grand sweep, enter the limits of the mill-yard.

"Our great log is rolled off the car on to the platform, and in his turn passes to the small car used for drawing logs up into the mill. A long rope attached to a drum in the mill is fastened to the car, and slowly, but surely, it travels up to the platform near the saw. Our log is too large to go at once to the double circular, hence the 'muley,' a long saw, similar to a cross-cut saw, only it is a rip saw, and stands perpendicular, must rip it in two in the middle to get it into such a size that the double circular can reach through it. This is rather a slow process, and as we have nearly thirty minutes on our hands while

REDWOOD BOLTS FOR SHINGLES being hauled to Arcata from Oley C. Hansen's Shake and Shingle Claim, (Ericson photo from California Redwood Assn.)

waiting for our log to pass through this saw, let us pay a visit to the shingle machine. This we find on a lower floor.

"The timber out of which shingles are made is cut into triangular or wedge-shaped pieces, about four feet long, and about sixteen inches in diameter. These are called "bolts." The first process is to saw them off into proper lengths. These blocks are then fastened into a rack, which passes by a saw, and as the rack passes back a ratchet is brought into requisition, which moves the bottom of the block in toward the saw, just the thickness of the thick end on the shingle and top end in to correspond with the thickness of the thin end. The block is then shoved past the saw, and a shingle is made, except that the edges are, of course, rough, and the two ends probably not at all of the same width. To remedy all this, the edge of the shingle is subjected to a trimmer, when it becomes a first-class shingle. They are packed into bunches, and are then ready for the market.

"We will now return to our log. It has just been run back on the carriage, and awaits further processes. A rope attached to a side drum is made fast to one-half of it, and it is soon lying on its back on the carriage in front of the double circular saws. Through this it passes in rapid rotation till it is sawed into broad slabs of the proper thickness to make the desired lumber. It is then passed along on rollers to the "pony" saw, when it is again cut in pieces of lumber of different sizes as required, such as two by four, four by four, four by six, etc. It is then piled upon a truck and wheeled into the yard, and piled up ready for the market. The other half of the log is sawed into boards, three-quarters of an inch thick. At the "pony" saw, part of it is ripped into boards, ten inches wide, and part into

plank, four inches wide. The boards, ten inches wide, pass along to a planing machine, and it comes out rustic siding. The four-inch plank passes through another planing machine, and comes out tongue and grooved ceiling.

"The heavy slabs which we saw come off the first and second time the saw passed through the log are cut into different lengths, and sawed into the right size for pickets. They are then passed through a planer, then through a picket-header, a machine with a series of revolving knives, which cut out the design of the picket-head the same as the different members of a moulding are cut out. Thus have we taken our readers through the entire process of converting the mighty forest monarchs into lumber. We hope we have succeeded in making the description of the process, in a small measure at least, as interesting to our readers as it was to us when, for the first time, we witnessed it. When you have witnessed the process of making lumber in one mill you have seen it in all, with the exception of here and there a minor detail. There are but few mills which use a "dummy" engine to draw their logs to the mill, most of them using horses or cattle on the tramways. The lumber and wood industries of this township will always make it of considerable importance and a prosperous future may reasonably be expected."

Duncan's Mill was not alone in the Sonoma redwoods. W. R. Miller had a mill nearby and in 1866, M. C. Meeker put one in operation south of Occidental. There was the Smith mill in Coleman Valley, and the Heald and Guerne Saw and Planing Mill Co. whose mill was carried away by a Russian River flood. J. C. Fiske was steam sawing lumber as early as 1860 and cut 42 million feet in fourteen years but his mill went into a decline after being sold to Fred Helmke. Fiske later had a shingle mill at Fisherman's Bay where nearby were the Platt Mill Co., the Clipper and Rutherford and Cooke's. The schooner *Lottie Collins* was in regular redwood trade between Stewart's Point and San Francisco. Said the magazine Wood and Iron in 1888:

Duncan's Mills Lumber Co. — a new one on this Coast — is composed of J. M. Dollar, William Fraser, and John Brennan. They came here from the East, and last fall purchased this tract with all its mills, railroad, logging and camp complement, from Mr. Duncan, through that well known and indefatigable friend of redwood, H. W. Plummer. This tract is admirably situated, directly on the line of the North Pacific Coast Railroad, whose track almost surrounds the mill, and then branches off into the canon towards what is now called Cazadero, but was formerly known as Ingrahams. This affords splendid facilities for logging, and several times a day the logging train runs out to the camps of the company, and the mammoth logs are brought into the mill.

The gentlemen composing this company are practical millmen; they have come here imbued with Eastern ideas, and are putting into practical use the knowledge years of experience has given them. That they have made a successful venture is already proved, and their output, which each year is bound to increase, is sure of a ready market at remunerative prices; for, no matter what the market price in San Francisco, they will have a home demand which will consume almost all that they can make for years. On their tract are some of the finest redwood trees that ever grew; towering hundreds of feet to the sky, their mammoth proportions dwarf completely the loftiest ideas of man, and he feels small and insignificant as he gazes upon these wonders of God's handicraft.

Across the Russian river from Duncan's Mill, and about four miles from there, is located what has been known as the Meeker tract. This is a tract of virgin forest, and now consists of about two thousand acres. This tract has been purchased by Messrs. William Westover, Robert Dollar, D. L. Westover and W. W. Westover, who, under the firm name of The Sonoma Lumber Company, intended to erect a mill, and proceed at once to cut the timber; but since our visit there the same gentlemen have purchased of Guerne & Murphy, their tract, lying north of the 2,500 acres, with all the mill facilities and railroad, so that at present no mill will be erected. This property, taken as a whole, cannot be excelled. It is tapped by the Donohue line of railroad, and the narrow-gauge is within a short distance of it. It is covered with a fine growth of trees, many of them measuring nineteen feet through, and the water supply is ample, so that the timber can be easily handled. On the timber tract is one of the finest groves of redwood on the Coast, and our artist sketched a scene on the spot where the Bohemian Club of San Francisco often congregate upon the occasion of their annual picnics. The picture is true to life.

This company will now proceed, at the Guerne & Murphy mill, to manufacture, and as fast as necessary, will develop the balance of their property, which is conceded to stand among the finest tracts on the Coast.

REDWOOD ROOST — "Rabbits live in trees . . . why can't we?" argued the man who boarded up the arch and took up residence in this Big Basin redwood, Santa Clara County. (Mrs. Hare photo from Bancroft Library, University of California)

From Massachusetts in 1849 came Dr. R. O. Tripp to the San Mateo region to establish a store and shipping center and to nurture a booming lumber camp at Woodside. In four years there were fifteen sawmills within five miles of the settlement sending redwood into San Francisco. Northwest of Woodside there were two Whipple mills, the Smith mill nearby and Horace Templeton's mill near Searsville. Bake and Burnham, operating a gangmill near Woodside, later moved it to Squealer Gulch.

By 1850 William Blackburn had logged off much of Branciforte Creek and Don Carlos Rousillon had cut timber from Sainswain's property, with R. G. Hinckley and John L. Selby, sawmilling on Soquel Creek. Then Judge J. H. Watson built a shingle mill on the Castro Grant near Watson-

BREEN BROTHERS OF CRESCENT CITY freighting the first Dolbeer donkey engine used in Del Norte County—headed east to Hobbs, Wall camp on what is now the Bertsch Tract. (Photo Ernie Coan Collection.)

ville and when he abandoned it, Rafael Castro worked it, finally leasing it to the Nichols Brothers —Joe, Ben and Milt. In 1851 a company of five deserters from a French ship built a mill on Aptos Creek, later selling it to Castro. This Spanish grandee also owned for a short time the mill on Soquel Creek formerly operated by Hanes and Daubenbiss.

In the Los Gatos country the Froment Mill was active on the site of the Forest House stage station and in 1851 Bill Campbell erected a mill just west of the present Saratoga. Then came the Moody brothers, James Howe, Wilson Webb, Stillman Thomas, McMillan and Bill Dougherty who started as a bullpuncher and rose to power in the Santa Cruz boom days with the Santa Clara Valley Mill and Lumber Co.

In 1859 a burned out sawmill was rebuilt by Capt. Grear on his Rancho Car da de Raymundo and the next year W. P. Morrison built a mill in Bear Gulch south of Woodside, moving it five years later to the headwaters of LaHonda Creek on the Pacific side of the mountains where it operated until 1872. Other pioneer redwood men here were John and George Moorem, Oakley, Pinchey,

Tuttle, Gardner, Spalding, Jim Gibbs and "Squealin' Sam" Aswell.

The western side of the redwood belt remained uncut for a longer period of time. Even when William Tuffley built a sawmill in 1852 on Pescadero Creek and another followed, the cutting was small and intermittent. W. Wadell's mill on Wadell's Creek in the '60's was one of the larger activities and farther up the coast Borden and Hatch carried on a thriving business started by Doolittle and Crumpecker in 1854.

What did these redwoods look like in those days? An account of a trip along the San Bruno Road into San Mateo County appeared in Hutching's magazine in August, 1860.

"Gently descending, the party saw a number of small cottages with near gardens in front, and which indicated their proximity to one of the many sawmills built in the woods for lumbering purposes. A little farther down stood the mill—a new one just finished—as in one night the former one, known as Jones and Company's, which cost $30,000 with 100,000 feet of lumber, worth $18 to $20 per thousand, were consumed by fire; the blackened ruins and burnt ironwork still lying in the old location, indicated the extent of the conflagration. Thus, in one night, the labor of years was swept away; alas! how many of these, and similar losses, there have been in this young State.

"The new mill, owned by Mills and Franklin, had just been completed and put in working order, and which is capable of cutting 15,000 to 20,000 feet a day. This with the produce of other mills, for several miles around, is conveyed on large and strong wagons to Redwood City, from whence it is mostly shipped to San Francisco, where it is wholesaled to lumbermen by the cargo at about $20 per thousand. The cost of getting the logs and manufacturing the lumber, averaging about $8 per thousand; transportation from the mill by ox teams to Redwood City about $5 per thousand; freight thence to San Francisco about $2.50; leaving about $4.50 per thousand for wear and tear on machinery and teams, interest on money invested, profits, losses and the general superintendence of the owners. Here Hawkins and Clary's patent regulator is used for gauging the lumber to any thickness required, in a moment, by which an immense saving of time is secured."

As in other redwood areas, the little mills faded out and combinations were set up. Whitehurst and Hodge had logged in the vicinity of Mt. Madrona Park during the '70's and '80's to supply their yard at Gilroy. The Bodish mill cut here also. By 1900 the Santa Clara Valley Mill and Lumber Co. was operating mills in Bear Creek, Newell Creek, Zayante and San Lorenzo River. Hubbard and Carmichael brothers of San Jose were cutting redwood on Oil Creek and Waterman Creek and on the China Grade adjacent to the Big Basin area, the colorful and dynamic I. T. Bloom was a leading producer. Another was H. L. Middleton's California Timber Co. F. A. Hihn, originally a penniless German peddler, founded a prosperous lumber business and the firms of Loma Prieta Co., San Vincente Lumber Co. and the Bloomquist family had big operations which were followed in later years by Monterey Bay Redwood Co. and Santa Cruz Lumber Co.

HOBBS-WALL: 80 YEAR SAGA

"In 1866 came Calib Hobbs and David Pomeroy of the firm Gilmore & Company Box Factory, San Francisco. The result of their investigation led to the establishment of the large mill and box factory in Elk Valley in 1866. J. E. Wall became associated with the organization which was afterwards known as the Hobbs-Wall company.

"There had been little thought given to the exporting of lumber, the principal business and enterprise of Crescent City and settlements being the forwarding of merchandise, through commission merchants. Supplies and provisions were shipped to the Crescent City port, brought in by steamers and an occasional sailing vessel, freight rates ranging from $10 to $12 a ton. There being no wharf, freight was unloaded onto lighters at a cost of

BERTSCH BROTHERS SHAKE MILL at Fort Dick, Del Norte County in the 1890s, seen dimly in the redwood forest. (Photo Ernie Coan Collection.)

ROLLING REDWOOD THE HARD WAY in the old days at Fort Bragg. It was slow work but usually sure—moving logs by jackscrews in 1890. (Photo Union Lumber Company Collection.)

$3 a ton and another dollar added for drayage to a warehouse. At that early date all merchandise for the interior was carried on pack mules, and it was not an uncommon sight to see from 150 to 600 mules being packed in one day. The scene changed with the coming of the wagon roads. Hobbs-Wall later rebuilt the Simpson & Wenger Mill on Lake Earl and operated it until the company went out of business.

"At the time of the original construction of the Lake Earl mill, some doubt arose as to the disposition of the lumber, how to ship and how to get it aboard vessels without too much handling. This problem was solved when Johnson contracted to build the first major wharf in the harbor for Wenger.

"The first shipment of lumber, made before the completion of the wharf, was rolled out to the lighters, the lumber having been hauled to the beach by oxen from the mill and piled above high water mark.

"The start of the century saw the lumber industry of Del Norte tied up with her first major strike of the loggers and mill men against the ailing wage scales of Hobbs-Wall Company. In early May of 1903, an agreement was reached in both their mills, Lake Earl and Elk Valley, and logging camps. The strike of only a few weeks had demoralized business. The terms of the settlement fixed

a minimum rate of $40 per month for unskilled labor. The best blacksmith was to receive $80 per month, as well as the carpenter for a ten hour working day. Head choppers got $60 per month, with board. The head sawyer was the best paid man on the job, $125.

"There was a great flurry of excitement at Crescent City in May of 1903, when Hobbs-Wall company had their first locomotive make its initial trip on the wharf track. It was, too, during this week, that the company closed a deal whereby they gained control and almost entire possession of the lumbering industry in the county.

"In 1909, the Hobbs-Wall Company were operating two sawmills in the county, several logging camps, their big store in Crescent City and other smaller ones at camps, twelve miles of railroad from their camps to the mills and the Crescent City & Smith River railroad from Smith River to their Crescent City wharf. Three hundred to four hundred men were employed, the Lake Earl mill having a ten-hour capacity of 40,000 feet and the Elk Valley mill at Crescent City, 100,000 feet. The company also owned and operated steam schooners, especially built for lumber transportation, plying between Crescent City and San Francisco and San Pedro."

—*from an article by Ernie Coan in Del Norte Triplicate Centennial Edition 1954.*

THE DAY OF THE BULL

Falling the big redwoods and getting the logs to the mill was not only hard work which had to be done by men who were used to working hard but it was always hazardous from all angles. A jackscrew could slip and a man crushed by the rolling log. He could be wiped out financially if the river flood washed out the booms and a year's cut floated out to sea. Work in the redwoods is graphically told in the account from the Humboldt Times reprinted in Pacific Coast Wood and Iron in 1888.

"Lumbering operations in Humboldt commence for the season usually soon after the Christmas and New Year holidays, providing it is what we call an open winter. But if a protracted rain sets in at that time, the beginning of operations is postponed until a clear spell intervenes.

"A crew of a dozen or more men are sent to each camp. This crew usually consists of a cook, several choppers, a few sawyers, and others to peel and ring the trees that are fallen. The choppers go in pairs, two men chopping at the same tree and cut. One of these, the head chopper, takes the lead, directing how the tree is to be felled, etc. Considerable responsibility rests upon the head chopper, as the amount of lumber obtained from some of the largest redwood trees and the safety of the adjacent timber de-

SWINGING LOG TO SKIDROAD in Albion woods, Mendocino County, in 1869. Dogger with maul stands atop log while bull whacker, with goad stick, stands ready to drive bulls into action. (Photo Pacific Lumber Company.)

pends largely upon how it is felled. The head chopper must then be a man of experience and judgment, as well as a good axman. Most of the timber is now felled with saws, but an undercut has first to be made, and the workmen must be handy with the axe as well as the saw. For this purpose 12-foot saws are generally used, which are worked by two men.

"First, the head chopper decides where the tree is to fall, marking out a path where it may fall with least injury to itself and other timber. On the steep hillsides it is usually uphill, as it will sustain the least injury in that way, but if the slope is not uniform, if the country is rugged, as much of it is, broken by sharp divides and abrupt depressions, then the choppers must use great circumspection in choosing a bed for the tree, as the immense weight of the redwood and its easy rift renders it peculiarly liable to injury in falling. Several thousand feet of the best lumber may be shattered and broken in falling one tree by an inexperienced and thoughtless hand. But the chopper has to look further than the safety of the tree which he is about to fall, for a dozen or twenty trees may be within reach, all of which are to be felled before any of the timber is removed; and it must ever be borne in mind that these other trees must have sleeping places reserved for them. A very good place may be found to fall one tree, but that may interfere with the falling of several others, which might be broken and partially destroyed by falling across its great trunk. So it will be seen that the head chopper has to consider well the nature of the ground in any certain cluster of trees, select with care the tree which may be fallen first with least interference with the others, and choose a bed for each tree with a map of all the trees as they will lie when felled in his mind before starting.

"The bed chosen, the next thing in order is to erect staging to bring the choppers on a common level and elevate them sufficiently to cut the bole of the tree above irregularities produced by the immense roots. The tree is usually cut six to ten feet above the ground, although the tendency at present is to cut them as low as possible in order to save lumber. When the staging is completed an undercut is made on the side of the tree toward which it is to fall. In making this cut both choppers work together, one of course being a right-handed chopper and the other left-handed.

"They use double-bitted axes, weighing from three to four pounds. The helves now generally used are the best second growth hickory, strong and elastic. These helves are manufactured mostly in Pennsylvania, specially for the trade, and cost here from 75 cents to $1.00 apiece. The helves are straight, and for chopping are from 38 to 42 inches long, which, in itself is a sufficient explanation why 3½-pound axes are commonly used. A man that swings over three pounds of steel all day at the end of a 42-inch handle must be possessed of much power and endurance. Strangers, even those familiar with lumber chopping in the East, might wonder why such long handles are used. It is all on account of the size of our trees. The woodman must be able to reach the center of the tree from his position on the platform, and it must be remembered that in many of our redwood trees the center is four or six feet from the outer edge.

"In making the undercut the utmost care is used in having the scarf extend on either side an equal distance

from the point toward which the tree is to fall. For this purpose what is called a 'gun' or pointer is used to indicate the direction in which it is to fall. So exact are all the preparations that a tree in falling seldom varies a foot from the line laid out for it.

"The undercut completed, the choppers commence with the felling saw on the opposite side of the tree and slightly above the bottom point of the cut. As they saw in steel wedges are driven in the crevice close to each other, and as the saw sinks deeper and deeper more wedges are used until the tree is forced bodily over by the mechanical power of the driven wedges, and falls when it is nearly severed.

"The next step in the process of lumber making is to cut the tree into logs. But first the sawyer must make a study of the tree to determine how it will cut to the best advantage. The logs are cut in lengths of even feet from twelve up to twenty feet in length. Where a cut is to be made a 'ring' is made in the bark, much as we used to girdle standing trees to cause their death. Then 'peelers' go on with flattened iron bars and take the bark off the tree. At some seasons of the year the bark removes quite easily, but peeling is, like all work in the redwoods, heavy and dangerous work.

"In sawing the tree up into logs an eight foot saw is used principally. It is handled by one man readily, but the sawyer must be able to file and set his own saw, which is a much more particular thing to do than draw it back and forth. Of recent years, however, in some camps where many sawyers are employed, a special man is employed to file the saws and keep them in cutting trim.

"As the days lengthen toward spring the crew is increased until it numbers forty to sixty men, and several million feet of logs are ready for hauling. The most serious, difficult and expensive part of the work connected with logging operations in Humboldt now commences. This is the construction of log roads. The matter of roadbuilding in the hill regions where the country is cut up into many ravines or gulches separated by sharp divides, and where a separate road must be built up each gulch to obtain the timber, involves an expense that would appall the novice. Sometimes several thousand dollars are expended on a road before a single log is reached. Where such a road has to be built up a ravine where but a limited quantity of timber exists, say a million or two million feet, it adds materially to the cost of logs.

"In connection with this item of road-making, many buyers of redwood timber land who are not familiar with logging in the redwoods, are liable to make a serious mistake. On a tract of 160 acres there may be 6,000,000 feet of standing timber, while the same amount of timber may be found on another tract of half the area. The uninitiated would naturally suppose that the 160 acre tract is the more valuable, because it contains more land. But if he does the logging himself, and expends $5,000 or more in making roads all over that 160 acres to get the 6,000,000 feet of timber, while on the 80 acre tract he might obtain the same amount of timber by an expenditure of $2,500 for roads, he will soon be undeceived. Redwood timber land increases in value, not according to the density of the timber, but in a greater ratio.

"But to return to the work in the woods, a crew of 'swampers' is started in the early spring to make logging roads. In later years they have had a valuable adjunct

BULL WHACKER WAS TOP HAND of woods crew. He got $100 a month for driving teams down slippery roads under goad and bellowed oath, keeping bulls shod, free from neck galls, horns properly capped. (Photo Escola Collection.)

BULLS ON BIG RIVER in Mendocino County. Eight and ten oxen were standard teams in the early days. Sometimes horses were used to lead, for quicker responses to bull puncher's urging. (Photo Escola Collection.)

in this work in the steam 'donkey' engine, invented by Mr. John Dolbeer, who has now sold his patents to Messrs. Marshultz & Cantrell, of San Francisco, who are manufacturing them. Nearly a hundred of them having been sold to different lumbermen and there can be no question as to their efficiency and that they greatly reduce the labor and therefore the cost of handling logs.

"This is a small engine such as is used for pile driving, or hoisting with an upright boiler, and a 'spool.' This engine is fixed upon a heavy bed or sled, and by the use of tackle and attaching blocks to stumps and trees, it pulls itself up the steep grades, and is taken into every logging camp and road. The engine is used to take old logs and stumps out of the way. But its principal use is to get the logs into the road for the team. As we have said, the 'donkey' usually goes to work about the same time as the 'swampers,' and by the time the rains are over and the roads ready for hauling, a large number of logs are ready for hauling. In 'swamping out' these logs and attaching them with 'dogs' and chains, they are 'sniped,' that is, the sharp corner or right angle of the forward end is rounded off, so that in the process of hauling it will not dig into the soil, or catch on any impediment. We are ready now for the team, and an additional force of men.

"As soon as the rains are over in the spring, the crew in the logging camp is increased. An additional force of choppers, sawyers, swampers, and 'chain tenders' come in with the team. The camp may now number anywhere from thirty to one hundred men, according to the amount of lumber to be gotten from that particular claim, or to the distance of the haul to get it out. The team is usually eight to twelve good strong oxen, or in some cases, six, eight or ten heavy horses. The team, whether composed of horses or oxen, is driven by a single man, who next to the cook is the most important personage in the camp, and receives by far the highest wages. And really the output of the camp is more dependent upon the teamster than upon any other man or set of men. With poor helpers the best teamster is handicapped, but experienced men are generally found for all the important positions, so that the summer's work depends upon the team and its driver.

"When the team starts with the load for the landing the work of the teamster and 'water packer' commences. This latter individual becomes for a time the arbiter of the life and fortunes of the team, teamster and load. It is his duty to accompany the team and keep the road generously sprinkled with water in front of the load. His carelessness may precipitate a load down a steep pitch onto the team, or it may 'hang the load up' in the most difficult place. Tanks of water are stationed along the

HOMER BAKER WAS BULL PUNCHER on this Salmon Creek (Mendocino County) skidding operation in 1883. Redwood logs were usually landed at creekside during spring and summer, floated to mills on high water in fall. (Photo Escola Collection.)

logging roadway and five-gallon coal oil cans provided with a stout wooden hand-piece are scattered along the road plenteously. These are usually filled with water on the way up the hill, for on the trip down everything is done with a rush, and there is no time to dabble in water. The 'water packer' or 'thrower' then must stay with the team however fast they go and supply enough water to make the logs glide smoothly with the least strain on the team. But woe to the 'water packer' who stumbles, or inadvertently spills a can of water at the brink of a steep grade? The logs, in that case, might take a sudden shoot and overwhelm the team before the teamster can give the word of command. Or if the supply of water is not sufficient at some hard pull then the load 'hangs up,' and men with jack screws have to work and get the logs started again. It will thus be seen that the 'water packer' must not only be agile, but he must use the utmost judgment in distributing the water. By a false step he may himself be thrown in front of and be run over by the logs, or by ill judgment he may drive the logs down onto the team and driver. This position, like that of teamster, is one of the most hazardous in the logging woods, and is also nearly as hard to fill satisfactorily. He, as well as the head 'chain-tender,' must work in harmony with the teamster, as it is absolutely necessary that the latter should have full say as to the size and make-up of his load.

"The teamster is a sort of boss in himself. In driving cattle he uses only words of command, reinforced by a 'goad stick.' This is a piece of crab-apple, yew, hickory or some other tough wood, about four feet long, with a brad in the end. This brad has to be occasionally driven in the thick hide of a lazy ox (and there is always one lazy ox in every team) to make him understand that promptness is a desirable commodity in the logging woods. But the main reliance of the teamster is a good pair of lungs, and what he lacks in vociferous shouting he makes up in frenzied gesture. The facility with which an ox-team works, however, depends much upon the leaders. With a pair of leaders that will supplement the driver's lungs everything is possible. Oxen in the logging works are driven from any point, and the driver may be on the 'off' side as often as the 'nigh.'

"The landing onto which the logs are hauled by the team is constructed usually of 'skids' or poles laid much in the fashion of the old 'corduroy' road, except that the poles are much larger, being ten inches or upwards through, and capable of sustaining the immense weight of the redwood. Usually there is sufficient room on the landing to hold several loads, or twenty or thirty logs. From the landing the logs are sometimes put upon cars, sometimes upon trucks and often directed into the water. That our readers may understand the full extent of the details of logging we will give the route of the logs in an instance where several methods of conveyance are used before they reach their destination.

"On the landing is a man whose duty it is to load them on the cars and keep the landing clear. In the case selected the car is simply a logging truck with grooved wheels and the railroad is a stretch of half a mile of nearly level road, and the rails are pine poles of the size that best fit the grooves in the car wheels. The propelling power is a team of four good horses. The car has a capacity of one large log, two medium sized logs or perhaps three small ones. A medium sized log is one from four to five feet through. The 'bunks' or bed pieces on the car on which the logs rest are of a fine-grained hardwood, either laurel, maple or oak, otherwise the great weight and general rough usage would soon splinter them up and wear them out. It is a fortunate season that they do not have to be renewed in mid-summer, even when the hardest and toughest material is employed.

"This car is on a level with the landing. The man who attends the latter, aided perhaps by the car teamster, rolls the logs onto the car by means of a jack-screw which is operated by crank. First, however, the 'chocks' are placed on the bunks on the opposite side of the car to prevent the log rolling too far. The load secured, the car departs, while the manipulator of the jack-screw gets one or more logs ready for the next trip.

"We had almost neglected to mention that the log is usually marked by the landing man. There are several reasons why logs should be marked. In the first place in the case under consideration there were three camps putting logs into the same stream for the same mill. Although the men in the three camps were working for the same proprietor, it is always desirable to keep accounts of the output of each, and thus estimate the cost of the logs. In the second place, the logs from this particular camp were taken from two different 'claims,' owned by different men, to each of whom the lumberman paid stumpage. Logs then must have one mark to designate the land from which they came, and as another camp was located on the same land as part of this one, a distinguishing mark for the camp also had to be made. But many of the logs may be 'sinkers,' and will get lost in the pond and not reach the mill for one or two seasons afterward. Now, as one of the camps is putting in logs by contract and takes the mill tally for the amount, a mark must be used to designate not only the 'claim' and the 'camp' from which they came, but also the year they were put in, so that this contractor may be sure of his credit. Thus there is often a complication of marks which might be confusing to one not familiar with the operations, but which the mill tally-man is sure to read correctly.

JACKSCREWS, SNATCH BLOCKS

Writing of Stewart, Hunter and Johnson's little operation in the Ten Mile River area, David Warren Ryder speaks familiarly of logging the hard way in "Memories Of The Mendocino Coast."

"The tools used at that time were very simple. The entire logging equipment for the night and day run of the mill consisted of two teams of bulls —each team five yoke—with the necessary chains and jackscrews, a few snatch blocks, some manila rope and of course, axes and cross cut saws. In speaking of this in later life, C. R. (Johnson) said he believed the entire logging equipment did not cost them over $5000. And this was to log 50,000 feet of logs a day — $100 worth of equipment for each 1000 feet per day produced. Today the investment would be fifteen times as great.

"Old timers will of course remember how logging was done sixty-five years ago, but those who never saw it will enjoy the vivid description C. R. gives in his memoirs. He writes:

" 'The jackscrew was an important tool in the woods. It was wonderful what two men, each with a jackscrew, could do to a log. They could get it out from a hole and turn it clear around. Besides assisting in getting it out, they were also used where the ground was very steep in jackscrewing down to where the bulls could get the logs. A good jackscrew as I remember it cost about $75. They were wonderful tools.

" 'The cattle also were wonderful. The bull puncher was the highest paid man the country had. A good driver got equal pay to the foreman. When he wanted his team to start a load, he would commence his antics; yelling at the bulls, and jumping up and down and hitting them with his goad stick until finally getting them all started and pulling together. And it was a pretty good load that they wouldn't start.

" 'Depending upon the size of the logs, a load would consist of from four to eight logs. Each log was snipped a little on the forward end, the head log being the largest, and decreasing in size as they went backward. The logs were fastened together with a short manila line having a 'dog' attached to each end. The dog was driven in through the logs.

" 'The sugler was another important man. He accompanied the load down to the landing, and his job was to throw water ahead of the load. For this purpose he had a long stick which he carried over his shoulder and a bucket attached to each end. From these buckets he threw water on the road just ahead of the load. Water barrels were located at convenient places so he could often replenish his supply. Where the road was very steep, not much water was needed. But in the comparatively level places, it was a great help — to make the logs slide over the wet places. Where the road was nearly level, skids were put in — sometimes running lengthwise with the road, sometimes across the road. Also, chains were attached by 'dogs' to the log, and on steep places were dropped, to act as brakes and prevent the logs from piling onto the bulls. The suglers were very expert at their work. They had to be nimble-footed, and quick in replenishing the water in their buckets and throwing it under the logs.

" 'After we had logged a few months we heard about a machine called the 'Dolbeer donkey,' and I went to San Francisco and got one. It consisted of an upright boiler with 6 x 8 engine, which was fastened onto the boiler near the top and operated the gear. The gear revolved an upright spool which in turn wound a manila rope which we fastened to the logs in order to yard the logs where a bull team could easily reach them. We used this machine only for yarding purposes. The logs were yarded about 150 feet. Where four to six logs had been yarded out, the bull team was hitched to them and they were hauled about a mile to the pond. We used manila rope entirely. Didn't know anything about wire lines until four or five years later. The introduction of wire lines was a great advance in logging. But we never did put lumber on a vessel any cheaper than we did then, in the early days of logging, with those crude tools.' "

BULLS IN KINGS CANYON

"The increasing 'settling up' of the valley," writes Lizzie McGee in the October bulletin of the Tulare County Historical Society, herself one of the settlers, "had created a need for lumber with which to build homes. A saw mill was here to thin out the suitable timber to supply the demand. Bull teams hauling logs from the woods to the mill over skid roads, picturesquely dotted the hill side. A bull team needed no harness. A yoke fitted across the necks of each pair of bulls, this hooked to a heavy chain that in turn was hooked to a big log, or sometimes several of them stretched out. The bulls were pretty securely tied together and lifted the load with their strong necks and shoulders. A span of six, eight or ten bulls represented a powerful lift. Bill McGee drove one team. He had Bright, Brigham, Buck, Brin, Star and Hank and others. With a sharp goad stick he commanded obedience. He got them in motion with a light jab on the rump of each one. They began to lean forward, backs humped, the yokes began to creak, the chains clinked and the logging chain straightened out. The bull whacker kept alert. If an animal didn't take a step when the rest did he got a good punch with the goad. It reached up, over and down on the rear of Mr. Bull. If the team was too reluctant to get in motion McGee managed to get in some quick action. In rapid succession he jabbed heavily each bull's rump and emphasized the jab with a stout swear word. They moved evenly into pulling strength. In the morning Bill's voice rang out loud and clear. 'Gee, Henry Bright, Brigham, Buck, Brin and Star' with his pet tuning up booster oaths that fascinated us children and horrified some of the more serious elders. By noon the volume of his vocal output was considerably quieted. A skid greaser and swamper accompanied each team. They swabbed on skid grease where going was tough. Bill McGee was killed when a log jumped and pinned him to a stump."

LOGS TOO BIG FOR BULLS in the Sierra Big Tree area were split with powder blasts—probably during Smith Comstock period. (Photo Harold G. Schutt Collection.)

LONG TURN WITH 14 BULLS under the goading of Bill Hargraves near Mendocino in '92. On steep grades, man ran with lead log, looping chain under it to hold it back. (Photo Escola Collection)

BULLWHACKING IN THE DEL NORTE COUNTRY

A writer in the Del Norte Triplicate Centennial Edition writes colorfully of bullwhacking in that area when the first mills were struggling to get quantities of redwood on the coastal schooners while the demand held strong in the building of San Francisco and the Bay cities.

"The dogwood springled the green of the slopes with their white blossoms. A team of six yokes of oxen were coming out of the heavy timber, slowly making their way down the skid-road into Smith River Valley, some of the oxen swinging their heads, some slapping their tails to discourage a not too particular horsefly, straining with their load of green logs. Nearer to the slough the long team moved with only an occasional crack of the long bullwhip in the hands of the bullwhacker.

"As the team moved closer the driver was recognized as Andy Hamilton, his prodding the oxen was the answer to the stillness that surrounded the moving load. Hamilton had the reputation of the one bulldriver in the Del Norte woods to get more out of his oxen without using a swear word.

"This in itself was an oddity in that most of all goaders could outcuss any class of men, be it sea, on the trail, in the saloon or in the woods.

"There was Frank Hussey, another bullwhacker of the Del Norte woods who, they say, had at one time the repute of having a louder, longer epithet personality of his own, with a wider range of bull-language repertoire, entirely outside the range of euphemistic oaths. Be that as it may, both Hussey and Hamilton had the reputation of good men, experts with the whip and getting their loads out of the woods and down the skid-roads to the boat landings in their own particular way.

"Andy and Frank and others of their ilk had made an art of driving bulls, lost art now. Very few loggers of today have ever seen a team of oxen at work on the muddy skid-road, not a hell-bent bullwhacker leap to the back of his wheelers and walk up to his leaders on the upper decks of the oxen; few have any idea of the swaddling days when it was bull-power that moved the logs in the Del Norte woods. Gone are those days of barn bosses, skid-greasers, hooktenders; the handskidder, with his own private maul which he guarded jealously and the skid-saddler who hewed half-moons into the skids that the logs might drag more true to the road.

DUST ROLLS UP as bulls road log in Ten Mile River area, Mendocino County. (Photo Escola Collection.)

NOON RESPITE from heavy labor. Two 14-ox teams water in creek at Gualala, one of the earliest logging opperations on the redwood coast. (Photo Union Lumber Company Collection.)

"Many of the bullwhackers graduated from the grease pot. The skid-greaser walked ahead of the logs, behind the bull team and daubed heavy, black grease on each skid that lay across the skid-road, six to ten feet apart—the thick grease lessened the friction of the heavy logs. Next importance to the bulldriver of the gang was his hook-tender who hooked the big logs together with their chains and dogs and took care of any snags and holdups and hedged round the hang-ups. The hook-tender's helper was called the hand-skidder and with his trusty iron-handled maul drove the dogs into the logs and pried them out at the landing with the iron handle and tossed the chains over the yokes of the oxen for the return trip back to the woods.

"The bullwhacker was the top hand of the crew. In the Del Norte woods a good goader received about $100 a month and found, working from dawn to dark, twelve to sixteen hours making up an average day. A good hook-tender drew down about $70 to $75 a month and found, and how those men could stow away the eats. The bull-whacker looked after his ox teams of five to eight yokes, saw that the beast's necks didn't become galled from the heavy rains and probably assisted the blacksmith in putting brass caps on the ends of the animals' horns and an occasional shoeing when on rocky ground.

"But among all their profanity, not so much from irreverence, but to release a pressure of inward steam, the hardy drivers, fallers and buckers, living a rough life, with hard work, among the sweat-permeated air, the bellowing of the oxen and the crashing of falling monarchs, were as a rule a loyal lot.

"Then came John Dolbeer of Eureka, who visioned a better way to handle the heavy logs than by bull power—put his small, new donkey engine to work in the redwoods which was the death knell of the brad-tipped goad and the rawhide-lashed bullwhip. The Dolbeer engine was a single cylinder, small vertical engine that powered an upright capstan, called the spool by the lumberjacks. Dolbeer first used heavy hemp rope, later came the steel cable. Even with breakdowns and occasional delays, with the additional crew of spooltender, firemen and engineer of the donkey crew, the Dolbeer engine proved a great improvement over the slower bull power with the bulls' frequent skirmishes with their one dreaded enemy, the yellow-jackets."

REDWOODS ON PARADE

The nation's first centennial was celebrated in Philadelphia in 1876. To Californians of the day it was notable in having a display of a simulated redwood tree and in having it met with c r i e s o f "Hoax!" There just wasn't any tree so big. Look where the pieces were joined together — proof that several small trees had been joined to fool the public.

The story behind this first prideful attempt of the redwood pioneers to scotch public skepticism about California's big trees is told in the Tulare County Historical Society's bulletin of October, 1950.

"What would permit the fair visitor to visualize these trees better than to exhibit a section of the trunk of a tree? A solid section couldn't be transported but the next best thing was done. Mrs. McGee reports, in 1875, Martin Vivian cut a big Sequoia near the General Grant with axes, a sixteen foot section was cut out and split into pie shaped pieces taking care to preserve the bark. They then split the heart out of each piece leaving a rim of bark and sapwood. These outer pieces were hauled out by Happy

WOE BETIDE THE LAGGARD bulls on grades like this who let the lead log bear down on their heels. Note action of puncher at right. Scene near Duncan's Mill, Sonoma County. (Mary W. Wisner photo Pacific Lumber Company Collection.)

Gap (near present Sequoia Lake) and Traver to Cross Creek and shipped to Philadelphia.

"Mrs. McGee's uncle, Israel Gamlin, had a squatter's timber claim in this area and he, with Mrs. McGee's father, Tom Gamlin, Poley Kanawyer and probably others, helped cut the tree. Vivian was not well-to-do and probably had some financial backing but there is no information about outside sponsors.

"Jesse Pattee says that Sam and Bill Harp, Huse Campbell and John Moore hauled the exhibit to the railroad. He recalls that when he first went to the mountains that it was a fad to fashion canes from sticks split from the centennial log.

"Park records confirm the general information above but mention no names. Since Vivian cut this tree on government land without permission tradition says that he was sentenced to a year in federal prison. Wallace Elliott in "History of Tulare County" (1883) says he was fined fifty dollars but should have been jailed for life for his vandalism."

And there were commercial ventures involving exhibition of the big trees. One is related in the files of the Farm Tribune (Porterville) and reprinted in the Society's bulletin of the same date.

"June 8, 1889. John McKiernan, of Cramer intends cutting a 26-foot in diameter redwood tree for exhibition sometime this month. This tree is situated near A. J. Doty's Mountain Home summer resort.

"August 3, 1889. John McKiernan felled the large tree at Mountain Home that he is to take to Europe, last Friday evening just at dusk. Many were disappointed as only two persons saw it fall. Many campers had gone from Summer Home to watch the work for the past two weeks.

"August 31, 1889. John McKiernan, of Pleasant Valley, informs that he will send down his big tree samples to Porterville next week. It is divided into eight distinct pieces which will be banded together when they arrive in Los Angeles where they will be shipped for exhibition.

"October 5, 1889. At last, the big tree, "California", which Messers McKiernan and Davidson have been cutting down in the Redwood forest above Frazier, is ready and will be under way to visit those places where anything from the "wild west" will be welcomed as a curiosity, and shortly, those narrow-minded sceptics who have never seen a genuine Giant of the Redwood groves will have to admit that the fabulous stories told of the world famous but little seen Sequoia gigantea are true.

" 'California' was cut from a tree growing in the Redwood grove home three quarters of a mile to the east of Frazier's mill and is a portion of a forest giant which grew to a height of some 300 feet and measured some 76 feet in circumferance at the base. It has been cut into eight separate pieces each weighing some 1,200 to 1,500 pounds.

"John McKiernan and three assistants accompanied "California" to Porterville, where they arrived at about

STURDY TEAMS HAUL REDWOOD TO MARKET from Love Creek sawmill near Ben Lomond, Santa Cruz County. (Photo from Boeckenoogen Collection.)

HARDY PIONEERS AND LUSTY BULLS logged the first redwoods—these in Santa Clara Valley. (Photo Boeckenoogen Collection.)

2:00 P.M. Tuesday. Monday morning will witness the departure of the stump for Visalia where it will be placed on exhibition for the forthcoming fair, after which it is bound for Tulare, Fresno, Merced, San Francisco, Sacramento, San Diego and San Bernardino. Eventually, it will bid a long farewell to its native state and will start for New Orleans via Texas."

Engaged in another enterprise of this kind, says a paper prepared in 1923 by Mrs. Jay Brown, were J. R. Hubbs and Ed Manley of Mountain Home. The tree, supposedly

for the Centennial Exposition, was cut off high above the ground, hollowed out, sawed into sections and "taken out via Happy Camp, Pine Springs, Rancherie and Mountain View. (ie the Kincaid Mill road). It was shipped from Tulare by rail to San Francisco where it was exhibited at Woodward Gardens and thence routed east to the Centennial. One by one the partners withdrew from the enterprise. First Manley, then McKiernan and before the tree left San Francisco Hubbs sold out for two thousand dollars, three thousand dollars less than the cost of preparation."

DOWN TO THE LANDING in Humboldt County—Pacific Lumber Company woods. 14-ox team on corduroy road. (Carpenter photo Pacific Lumber Company Collection.)

PLACID BULLS PAUSE FOR PICTURE but can't work up many smiles. A good bull puncher took care of his charges since they made a living for him but he got work out of them with two persuaders—a brad-tipped hickory or yew goad stick and his own lusty lungs. With a good pair of leader bulls and a stout bellow, the puncher could do wonders with any ox team, the lazy animals jogged into action with a jab of the stick, the steel point penetrating the hide. (Photo Boeckenoogen Collection.)

8 HORSES 2 BULLS on this Big River road. Animals just in front of lead log were called "chainers." Bull whacker here is Bill Ross. (Photo Escola Collection.)

LANDING AT GREENWOOD—early 1900s. Bull have been unchained from logs for return up skidroad. Horizontal spool donkey will load logs on narrow gauge cars. (Photo Union Lumber Company Collection.)

MUSCLE MEN FALL REDWOOD at head of Sonoma County skidroad. Note horses and side hitches on logs to bring them into line on road. (Photo from Pacific Lumber Company Collection.)

BULLS MEET TRAIN at Gualala landing. Logs were rolled on small "bobbie cars" by peavies and jackscrews, secured by chains and wood blocks. (Photo Escola Collection.)

TWO BULL TEAMS arrive at Hook's Mill in Santa Cruz area. (Photo Boeckenoogen Collection.)

FIVE-YOKE TEAM OF OXEN pulling redwood over skidroad to Bertsch Shingle Mill at Fort Dick, Del Norte County. In foreground with goad is Will Tryon, well known bull puncher of the area. (Photo Ernie Coan Collection.)

MENDOCINO ON THE MOVE

With the Meiggs-Ford-Williams advance into Big River in 1852, the coast redwoods found their first great utility. The rash of fires in San Francisco kept the demand for lumber steady and the Mendocino Saw Mills continued to supply a big portion of it. Harry Meiggs, in deep financial difficulties, was forced to leave the country and the other partners reorganized and worked out of debt. In 1863 the Big River mill burned and was rebuilt. a mile of track laid over which oxen pulled the cars of lumber to the schooner loading chute on the Point. The port of Mendocino was established here.

During these ten years the Albion River mill was built and passed into the hands of pioneer A. W. McPherson. The mill burned shortly after but the canny Scot with partner Henry Wetherby rebuilt it in 1867. Their logging boss, N. E. Hoak, had seventy-five men and four yoke of oxen in the timber. Later a railroad was built in the Albion gulch to bring logs to the mill.

Smaller mills were built at Gualala by John Rutherford and George Webber, at Hardscratch by Tift and Pound (water-power), on the Little River by Charles Pullen for Silas Coombs and Ruel Stickney, on the Navarro River by Tichnor and Hendy, on the Albion by A. G. Dallas and on Caspar Creek by Kelly and Randall. There were inland mills at Calpella (Thomas Elliott), Ukiah (Stephen Holden), Sherwood Valley (H. T. Hatch), Ackerman Creek (E. Prior), Cavelo (Andrew Grey), Laytonville (G. F. Bennett) and Willits (the Blosser brothers).

The Caspar Lumber Company had a long and stirring career. In 1864 Jacob G. Jackson bought out Kelly and Randle and under the name Caspar River Mills expanded and prospered. Additional owners in 1880 were F. A. Wilkins, Henry Fisher and Charles G. Jackson, all of San Francisco, and E. Sweet of Santa Cruz. In 1889 the mill burned but was rebuilt at once. Two years later J. G. Jackson died and his daughter, Mrs. Annie E. Krebs, took over the management. Later this passed into the hands of her sons C. E. De Kamp and C. J. Wood, who controlled the company until it stopped operating in 1946. James W. Lilley was manager in the latter years, having worked in the Caspar woods and mill since 1918.

The Gualala mill was another pioneer enterprise which lived to successful old age. In 1868 William Heywood and S. H. Harmon purchased a half interest in the Rutherford and Webber mill. Then

THE GANG MILL of Noyo Lumber Company. Lumber was chuted to barge and then loaded on schooners. (Watkins photo from Bancroft Library, University of California.)

FALLING BIG TWINS in Rockport woods about 1910. Biggest tree was 23 feet in diameter—the undercut a week's heavy work. (Photo Rockport Redwood Collection.)

in '72 the railroad was extended to Burns Landing, 2½ miles south of the mill, and in 1903 the Empire Redwood Company purchased the business, the mill burning three years later.

In 1914 E. B. Salsig, who had acquired the timber supply, started to build a new mill on the Gualala. The following year this unfinished building became the property of American Redwood Company. The owners, James G. and Robert Long, were mining men, having been driven out of Mexico by Pancho Villa. They completed the mill but sold out in 1920 to National Redwood Company. After lying idle for seventeen years it had brief usefulness as the Gualala Lumber Company, part of that time under lease by the Southern Redwood Company, and was dismantled in 1943. The Empire Redwood Company was now reorganized and in 1946 began operating once more only to have fire destroy the plant the next year. It was then rebuilt and operated under lease to the Al Boldt Lumber Company.

In 1859 A. W. McPherson, part owner of the Noyo River mill, acquired the rights to build a wharf and schooner moorings at Navarro. The sawmill was built in 1861 by Tichnor and Hendy, the latter of which two years later sold out and the firm became H. B. Tichnor and Company. Robert Byxbee acquired an interest and when Tichnor died in '83, Byxbee merged his ownership with Joseph Clark to form the Navarro Lumber Company. In 1886 the mill burned and was rebuilt farther up the river. The panic years of 1893 to 1895 ruined the business and Byxbee bought it at sheriff's sale but the mill remained idle and fell into disrepair. It burned in 1902, which ended a spotted career of forty years, and the buildings were sold to C. R. Johnson and his associates.

The Littleriver mill which Coombs and Stickney had operated since 1864 changed ownership ten years later with C. A. Perkins acquiring Stickney's interest. A new mill was erected at Buckhorn with a wooden railroad carrying lumber to a chute on the south side of the harbor, both mills cutting until the timber was depleted.

FAMOUS ROCKPORT SPAN—Steel wire suspension bridge 275 feet long from mainland to Flattop Rock was built in 1876 and used by Cottoneva Lumber Company, New York and Pennsylvania Lumber Company and Southern Redwood Company under Finkbine and Guild. (Below) Store and boarding house in 1890. (Photos Rockport Redwood Company Collection.)

J. H. WONDERLY STARTED USAL MILL (above) on Usal Creek above Rockport in 1889. Five years later it was purchased by Robert Dollar who built the schooner *Newsboy* to ship his lumber and went on to become a shipping tycoon. (Center) Navarro Lumber Company mill on south side of Navarro River. Lumber was barged out into harbor and loaded on schooners. (Right) Sawmill at Duffey, between Fort Bragg and Willits. (Photos Union Lumber Company Collection.) (Below) Wages Creek Mill near Westport. (Photo Escola Collection.)

The Garcia River mill had been built by Stevens and Whitmore who sold out in 1870 to Byron Nickerson and Samuel Baker. After twenty years in their hands, the mill passed to L. E. White who operated it until a fire destroyed it in 1894. Ben Severance was woods boss part of this period.

Eight miles from the ocean and town of Whitesboro on Salmon Creek, a sawmill was bought in 1874 by James Pullen and Grey, the latter president and general manager, and operated until 1900. A narrow gauge railroad brought lumber

down to Whitesboro where it was loaded on schooners for San Francisco and San Pedro. Brett and White had a mill a mile above the creek mouth and this became the L. E. White Lumber Company. It operated the tug *Aetna* and its railroad ran three-quarters of a mile up the creek and then to the creek drainage at Pullen's mill.

W. R. Miller, in 1877, built a mill at the mouth of Cottoneva Creek and two years later erected a wharf and chute that excelled any on the coast. It included a steel wire suspension span 275 feet to a cable tower on an island. In 1887 this operation was taken over by the Cottoneva Lumber Company but with timber depleted and fire destroying the mill in 1900, the business ended. In 1907 the New York and Pennsylvania Lumber Company purchased it and the Hooper mill on Hardy Creek, only to have its holdings taken back by the Cottoneva firm. The mill was destroyed by fire in 1912.

A new, modern, complete electric mill was built at Rockport in 1924-26 by a Mississippi firm, Finkbine-Guild, who had purchased the property. These new owners modernized the town and built the first large family homes. They constructed railroads for logging and handling lumber. Locomotives, railroad cars, and railroad steel, all were brought in by ship and trucked over tortuous roads. They constructed a suspension bridge from the mainland to a flat top rock, in the harbor, from which they loaded vessels with timber cants. These were waterborne to Jackson, Mississippi.

In 1928, the property passed into the hands of the Southern Redwood Corporation. They closed

the mill in 1929 and it remained closed until the Rockport Redwood Company was formed by Ralph M. Rounds of Wichita, Kansas, in 1938 and purchased the property.

The suspension bridge had fallen down during the period of closure, 1929-1938. In order to load vessels with lumber, a skyline between the mainland and the flat top rock was erected. By sling, the lumber would be swung high over the water to the rock and from there, by ships' cable, to the vessel. After two ships were loaded out, this was abandoned in favor of lumber truck hauling to the railroad in Fort Bragg, or out to Highway 101, and thence to California customers.

Tractors and trucks took over when, in 1942, fire destroyed the installation. The Greenwood (Lawson) flyer, slack line skidder, was moved from the woods to a permanent location at the long pond, as a decking machine. One of the original machines of this kind and possibly the last to operate, it handled as much as 12-million feet of logs a year.

During the latter years of the century other lumbering activity on the Mendocino coast began in districts where older firms had failed. The newer mills made gallant efforts to survive but with market conditions uncertain, manufacturing more complicated and costly, consolidation began to take place which gave more strength to the strong and dealt killing blows to the small independents.

The Pudding Creek Lumber Company, founded in 1886 by Capt. Samuel Blair, a retired sea captain, with Alex McCallum, made solid progress to 1900 under the name, Glen Blair Redwood Com-

pany. By this time the Union Lumber Company was growing toward eventual domination of the Mendocino redwood production and acquired one-half interest. The mill was rebuilt in 1908 and continued for twenty more years.

The Howard Creek mill, two miles north of Westport, had been built in 1875 by McFaul and Williams and later passed into the hands of the West Coast Lumber Company. It went into bankruptcy in 1914. The Wages Creek operation, started in 1881 by Pollard and Blaisdell, was also near Westport. It failed the next year and Gill, Gordon and McPhee ran it for several years. In 1889 it was moved to De Haven on Gordon's ranch and became the Pollard Lumber Company. A second Wages Creek mill was operated in 1881 by W. Graham and partners Chester and McGowan. It failed four years later and Hanson Hilton acquired it, the name changed to the California Lumber Company.

The Usal Redwood Company, with J. H. Wonderly as president — 1889 to 1891 — had a mill, 1600 foot wharf and three miles of railroad at Usal. In 1894 Capt. Robert Dollar became owner and his first steamer, the *Newsboy*, was used in the Usal to San Francisco trade. In 1902 the mill was shut down and fire swept the community including the mill. It was never rebuilt.

The Wendling Milling and Lumber Company, opened a sawmill plant fifteen miles up the Navarro River in 1903. The Stearns Lumber Company — A. G. Stearns, president — purchased it in 1905 and Standish and Hickey in 1913 when it became the Navarro Lumber Company. It was sold

in 1920 to Albion Lumber Company. This constituted a second sawmill for the Southern Pacific Land Company.

Inland the Northwestern Redwood Company mill, two miles west of Willits, which had been built in 1901 by A. W. Foster who owned the Willits Hotel, a mercantile building and lumber yard, closed down in 1926 and was taken over by Irvine and Muir. These men had a mill on the Noyo River in 1909 at the settlement of Irmulco, also one in Two Rock Valley. Both mills closed down in 1912 but the latter started sawing again in 1916. Ten years later Irvine and Muir took over the Northwestern but stopped operation in 1928.

Other mills active during this period were A. Haun and Sons at Branscomb, built in 1884; Wehrspon and Ornbaum, starting in 1896; Alpine Lumber Company east of Fort Bragg, 1902; Ukiah Redwood Company and mills in Anderson Valley, Potter Valley, Covela and Laytonville.

There were latter day redwood operations near Branscomb — McDougall Lumber Company which eventually became Wilson-Beedy Lumber Company which in turn was taken over by its mill superintendent, Vernie Jack; Ben Mast's sawmill near Laytonville; W. C. Thompson's mill operated by Crawford Lumber Company; Harold Casteel mill at Willits which after burning was rebuilt by Pacific Coast Company; Hollow Tree Lumber Company near Hale's Grove, sold to D. M. W. Lumber Company, but took over another mill near Ukiah; Jensen Lumber Company, Willits, sold to Little Lake Lumber Company; Wolf Creek Lum-

ber Company at Jackass Creek; Coombs Lum Company, south of Piercy; Ridgewood Lum Company at Willits; H. E. Casteel Industries Willits; Aborigine Lumber Company near F Bragg; Mendocino Wood Products at Ridgew Ranch.

TAN BARK LANDING (opposite) near Fort Bragg in 1912. Ph on next page show sequence in barking oak trees which is spersed redwoods in this area. (Photo Union Lumber Com Collection.)

YARDING IN THE 1900s with upright spool Dolbeer donkey. ged together, the logs were then pulled down hill to skid and then to river landing by bull donkey. (Photo Union Lu Company Collection.)

PEELING TAN BARK was substantial means of existence in Mendocino woods in early days of the century. (Above) peeling after oak tree had been felled, (left) before falling. (Right) Dry chute on hillside to bring bark to landing. (Photos Union Lumber Company Collection.)

IN UNION THERE IS STRENGTH

For an hour the redheaded youth watched the bulls skidding the big logs into the creek, groaning and snorting under the violent goads of the puncher. At the mill he watched the boards fall away from the saw and drop to the piles where the yard men swung them up on the heavy wagon. A six-horse team hauled it away and he walked with the driver as the load swayed and strained at the stakes. Five miles over the rough road and they came to the rocky cliff where a spindly chute led down to the sea that boiled around the rocks.

That was in the winter of 1881 and the youth was C. R. Johnson at Stewart and Hunter's saw-milling enterprise at Mill Creek and Newport. Before the next year was out he would be a partner and the nucleus of the Union Lumber Company, the most expansive redwood business in Mendocino County and one of the "big three" in California, would be formed.

Calvin Stewart, covered wagon pioneer, had come from the Sonoma coast to the Ten Mile River area and established a shipping point at Bridgeport Landing, one at Newport three years later. By this time his brother-in-law, James Hun-

ter, had come north from Vallejo as partner. They built the sawmill on Mill Creek and a loading chute down the Newport cliffs to the little "dog hole" where small schooners could anchor and load a piece at a time.

Determined to work and live in California, C. R. Johnson, son of a Michigan lumberman, could not find what he wanted in San Francisco. He liked Stewart and Hunter and their small business and found they were eager to take in a third partner. Johnson returned to Michigan full of facts, figures and enthusiasm, obtaining from his father the necessary capital for the venture. In December, 1882 the firm became Stewart, Hunter and Johnson.

The first year, with a night shift operating, the mill did well but C. R. wanted more production. Mill Creek was not suitable for a second mill and Newport had no qualifications as a shipping point. The level flat of Fort Bragg did and would accommodate a mill of any size. By personally sounding the bottom he decided a wharf could be built behind the protection of reef rocks where vessels could be loaded "alongside" instead of "off shore."

Again Johnson went east and after being ridiculed for having the audacity to imagine profits

FIRST FORT BRAGG SAWMILL built in 1885 by Fort Bragg Redwood Company. Mill burned in 1888 and was replaced. In 1891 this firm merged with Noyo Lumber Company to form Union Lumber Company. (Photo Union Lumber Company Collection.)

NINE TWENTY-FOUR FOOT SAW LOGS were cut from this 12½-foot in diameter redwood in Union Lumber Company's woods near Fort Bragg. (Frederick photo from California Redwood Association.)

out of "those so-called mammoth trees," his father and two friends, Michigan Senator Stockbridge and James L. Houghteling, took stock in the company — Fort Bragg Redwood Company.

The first successive steps were — buying out Stewart, Hunter and Johnson, securing large tracts of timber from McPherson and Wetherby back of Fort Bragg, on Pudding Creek and along the Noyo River. By 1884 all negotiations were completed, the new mill construction started, new machinery ordered and some moved up from Mill Creek.

Piling was purchased from White and Plummer on the lower Noyo but McPherson and Wetherby placed an armed guard on their property to prevent any trespassing. So C. R. Johnson and Dan Corey, hs timber cruiser, towed it by rowboat, two sticks at a time. Building the wharf was the biggest undertaking the Mendocino coast had seen and it was beset by many troubles. It was finally finished and on the morning of November 16, 1885, the mill whistle sounded a long blast. What was destined to be one of the three largest redwood projects in the world was under way.

M. J. Sullivan was the first sawyer. Tom Johnson came a few years later and stayed fifty years. Fred Johnson was the man who created the wharf, aided by Luke Maddux. Two millwrights, John Cummings and Jack Ross, bossed the building of the mill which used hewn timbers. And among the Mill Creek crew to come to Fort Bragg were Charlie Freeburg, Chris Beck, cruiser Dan Corey, Charlie Banker and Alex Sanderson.

C. R. Johnson had heard of band saws beginning to replace circulars in the middle west pine states. He procured one supposedly made for the big redwoods. But the steel was not tempered right, teeth not spaced or pitched properly and it kept running off the wheel. It took months to conquer this trouble but then, operating twenty-four hours a day, production went over a hundred thousand feet in those two shifts.

For two and a half years the mill hummed and then — fire. There was a sudden explosion and Henry Little, down in the engine room, saw flames streaking up the thick walls. He quickly connected a pump, then ran to sound the whistle. As he yanked the cord, it fell limp. The fire had won.

With inadequate insurance the disaster was crippling but Johnson went to work immediately to work raising money for rehabilitation. He obtained this, a new mill was built, only to meet head on, in 1891, a new set of obstacles. The first effects of the business depression of the '90s were being felt and the Pudding Creek timber was almost exhausted.

Logs could be cut on the Noyo and brought down by the spring freshets but it would mean damming the river for storage, and piers set in the mouth to hold the logs from flooding out into the sea. Another alternative was to build an expensive railroad.

The need for more money to accomplish either of these projects, gave birth to the Union Lumber

STEAM SCHOONER WHITESBORO loading at L. E. White Company's wire chute at Greenwood. (Photo Union Lumber Company Collection.)

GREENWOOD INCLINE — (Below left) Horse hauling load of lumber to wire chute shown above, 1890. (Right) Chute as seen from steamer deck. Horsedrawn cars came along track on shelf from left. (Photo Union Lumber Company Collection.)

OLD APRON CHUTE owned by Garcia Mill Company at Point Arena. Beyond chute is loading platform of Tindall, Kimball and Wille. It did not extend to land and was destroyed by storm shortly after this picture was made. (Photo Union Lumber Company Collection.)

Company. C. R. Johnson persuaded W. P. Plummer and C. L. White to merge their interests with his and a new company was formed on August 17, 1891. It was daring and costly but plans were laid to extend the railroad through the mountains between Pudding Creek and the Noyo watershed by means of a tunnel. This would tap a 25 year timber supply and be the first link in a mainline rail connection.

It was hard to get experienced men for tunnel work. Chinese were brought in, the white population was incited to violence and when the Orientals were driven south to Mendocino, Sheriff Standly forced a stand against the rioters, persuaded them to allow the Chinese to finish a job they would not do themselves. The tunnel was finished.

In the face of other obstacles, the lumber market was lethargic and Johnson went to San Francisco to force the sale of Union products, leaving Plummer in charge of mill production. He made frequent trips to Los Angeles, sold lumber to the Southern Pacific Railroad, C. A. Hooper and great quantities to Newport Wharf and Lumber Company, a line yard concern at Santa Ana. Through this energetic campaigning, strict economy and careful management, Union kept going where most of the other mills shut down.

Facilities of Union Lumber Company were improved. The wharf and railroad were extended and the National Steamship Company formed to control water shipments to California and handle passengers and general freight between ports. Also the company obtained a half interest in the Glen Blair Lumber Company, established in 1886 by Capt. Samuel Blair and Alex McCallum.

And so the thriving Union enterprise faced 1906. "Everything seemed set for a record year," says David Warren Ryder in his book, Memories Of The Mendocino Coast, "then on that never-to-be-forgotten morning of April 18th, the earth rocked so violently it seemed some giant had taken it in his hands and was shaking it as a cat shakes a rat. When it was all over, Union's mill was off its foundations and badly wrecked, and a large part of Fort Bragg was destroyed. As in San Francisco, the quake itself was bad, but the fire which followed was vastly worse and did most of the damage. In his memoirs, C. R., who happened to be in Fort Bragg at the time, presents a vivid picture of the events of that fearful morning:

"'The quake wakened me but I hardly had time to realize what was happening before our mill superintendent, H. C. Johnson, came rushing into my room and said

(Continued on page 78)

THE WATKINS PICTURES

"In the spring of 1956 Mr. Eugene Compton, an associate of Professor J. W. Johnson of the Division of Mechanical Engineering, made a surprising discovery of a portfolio of pictures which at his suggestion the owner gave to the Bancroft Library. The pictures are the work of one of the most noted photographers in the United States in the period after the Gold Rush, Carleton Emmons Watkins, who in 1868 won the first prize at the Paris International Exposition awarded to the United States by the Committee on Photographic Landscapes.

"The 53 photographs cover the very earliest lumbering activities on the Mendocino Coast, in which Mr. J. B. Ford, the grandfather of the donor, Mrs. Lewis Pierce, was an outstanding pioneer. Watkins' photographs, which measure approximately 16″ x 20″ in size, are magnificent specimens of the photographer's art. This portfolio, which is in almost mint condition, pictures the sites where the first lumber mills were erected, the rocky seacoast, Indian villages, and other early views of historic value. The picture of Little River before a mill was constructed there is of particular interest, because the cycle of this spot is now complete, from primitive forest through mill and mill pond, thriving lumbering community with wharf and shipping point, to the creation of the excellent and widely known Van Damme State Park, which preserves this as a forested area. Little evidence of this historic chain of events now remains except through the medium of photographs.

"J. B. Ford, in partnership with Meiggs and Williams, built the first mill at Mendocino City in 1852 to furnish lumber for San Francisco, which was then expanding very rapidly, and thereby became the founder of Mendocino County's lumber industry, which has thrived for more than a hundred years. This portfolio of pictures is from a personal collection of Ford, to whose granddaughter, Mrs. Pierce, the Library expresses its appreciation for this gift, which makes the collection of Watkins photographs outstanding." — *from Bancroftiana, publication of the Bancroft Library, University of California.*

SETTLERS' HOMES ON SAND BAR at outlet of Caspar Creek. Schooners stand out to sea waiting for lumber from mill upstream at right. (Carleton Emmons Watkins photo from Bancroft Library, University of California.)

CASPAR MILL IN 1865 when operated by J. G. Jackson who had purchased it from Kelly and Randall. As Caspar River Mills, operation was one of longest in the redwoods—1864 to 1955. (Carleton Emmons Watkins photo from Bancroft Library, University of California.)

there had been a terrible quake. I hastily threw on some clothes and went down to the mill, which was a good deal of a wreck. The mill building had an angle of about twenty degrees. The smokestacks had fallen down, the furnaces were down too, sparks were coming out of the furnaces and a fire was imminent. The firemen reported that the pipe connections were all broken and there was no chance to get water.

" 'Luckily there was a locomotive under steam, so we hastily summoned it and ran it down the track, which at that time was laid on the pond dam, and so got the locomotive close to the power house. By this time there was a big crowd of men around the mill and they connected the locomotive boiler with the fire pump which could get water from the mill pond. We did this in remarkably quick time and by pouring water into the furnaces put out their fires and removed all danger from the mill.

" 'Meanwhile fires had broken out uptown and Capt. Hammar of the steamer *National City,* which was lying

alongside the wharf at the time, came up to the mill with some sailors and gathered together all the hose he could find. The water pipes of the town were broken and no water was available. The hotel and several other business buildings were already on fire. Captain Hammar ran his hose up and got water on the burning buildings.'

"As a result of Captain Hammar's action the fire was checked and part of Fort Bragg saved. But, like San Francisco, much of it burned down, and many people were left homeless and with only the clothes they had on their backs. However, disaster and distress bring out the good in most of us, and the people of Fort Bragg helped one another. Those who had food and clothing shared them with those who did not. From the Company store, C. R. gave out food, clothing, and blankets as long as the stock lasted. And to those who needed lumber for rebuilding he supplied it with the understanding that they could pay for it when they were able. He rushed repairs on the mill so that it could begin providing employment

C. R. JOHNSON WOKE UP TO THIS on that fatal morning in 1906. The earthquake wrecked Union's sawmill and most of the town of Fort Bragg. It took three months to get sawing again. (Photo Union Lumber Company Collection.)

ALBION IN 1917—The first mill here was water-powered, the second McPherson and Wetherby's steam mill in 1854 which burned in 1867, was immediately rebuilt. (Photo Escola Collection.)

WHITESBORO ON SALMON CREEK with L. E. White Lumber Company's mill smoking up the gulch. About 12 miles of narrow gauge railway ran up to Pullen's Mill. White mill ceased operations in 1892, was dismantled in 1898 and machinery moved to Greenwood. (Photo Union Lumber Company Collection.)

again as soon as possible, and in encouraging his fellow-townsmen to rebuild their stricken city, he went so far as to instruct the wholesale supply houses with whom he dealt in Oakland and San Francisco to restock a rival Fort Bragg merchant and charge the goods to Union Lumber's account.

"After the fire was out in Fort Bragg, and he had given instructions for the repair of the mill, C. R. took the steamer *National City* and went to San Francisco. There he found all the ruin and destruction he had seen in Fort Bragg multiplied a thousand fold. The scene was one of great confusion, but C. R. and Captain Hammar managed to help a little. There were a lot of invalids who were to be evacuated, and the *National City* made several trips to transport them all to Oakland. While in Oakland on their last trip, C. R. bought supplies for the

steamer and for Fort Bragg, and the next day the ship sailed for Fort Bragg, taking up a number of Fort Bragg and Mendocino people who had happened to be in San Francisco at the time.

"Repairs to the mill proceeded rapidly and in about three months it was in operation. Demands for lumber to rebuild San Francisco, San Jose, Santa Rosa and the smaller cities which had burned, kept Union operating at full capacity, and the employment this provided—at the mill, in the woods, and on the wharf—materially helped and hastened Fort Bragg's own rebuilding."

By this time the properties of the company included the Mill Creek operation, the White and Plummer lumber and tie business, the Glen Blair mill, Little Valley Lumber Company and a most

FERRY ACROSS THE ALBION RIVER—at extreme left. Mill is up river to left and workers lived in houses shown on sand flat. (Carleton Emmons Watkins photo from Bancroft Library, Uuniversity of California.)

important addition — the big Mendocino Lumber Company, latter day development of the Mendocino Saw Mills of Ford, Meiggs, Williams and Lansing. With it came much able personnel including John S. Ross, Sr. The Mendocino mill continued in production until 1937, shutting down after cutting three and a half million feet of fir from a salvaged Benson raft from the Columbia River.

After the earthquake, C. R. Johnson and associates organized the California Western Railroad and Navigation Company and in 1911 the first train ran from Fort Bragg to Willits. A few years later the Ten Mile River railroad opened up a vast supply of timber in that area. After 1920 the company instituted a reforestation program with a large nursery. Fire, drought and other elements destroyed the young trees but improved logging conditions after 1930 aided this program and made new forests possible. With carefully managed fire protection they tend toward a perpetual yield of redwood timber.

In 1939, after Union had ridden out the financial storms of depression, and C. R. Johnson was eighty years of age, he resigned from the presidency, was succeeded by his son Otis R. Johnson. In this same year Union suffered another disastrous fire in the planing mill and dry storage sheds. In rebuilding, substantial improvements were made for economy and efficiency of operation, all adding

to growth and prosperity. With his father's death in 1957, C. Russell Johnson became Union's president in its 75th year of progress.

WOODS BOSS-INVENTOR — At 18, from England via the South Seas and Canada, came Harry H. Holmes to work for Union Lumber Company and help build historic tunnel No. 1. He became Union's logging superintendent in the late 90's and remained so for 30 years. Holmes invented the donkey fairleader manufactured by W. H. Worden Co. and pioneered use of logging inclines. (Photo Union Lumber Company Collection.)

COMPOUND ENGINE for skyline logging—Smith and Watson 13"x13" 4 drum machine used at Camp 22 on South Fork of Ten Mile River. *(*Photo Union Lumber Company Collection.*)*

WHITESBORO ON SALMON CREEK —mill operated by L. E. White Lumber Company. At extreme left, partly hidden by hill, is Barton Hotel; other buildings near are company store, post office with dance hall on second floor, roundhouse. (Photo Escola Collection.)

SIXTY-INCH REDWOOD CANT going through ten-foot bandrig, one of three bandmills at Union's Fort Bragg plant. (Photo Union Lumber Company Collection.)

FIRST STEEL WHEEL on band saw in redwoods. In the 90's C. R. Johnson, founder of Union Lumber Company, braved ridicule in introducing band saws which first used wooden wheels. Arm at left controlled belt to wheel, was hinged for clearance of big redwood logs. (Photo Union Lumber Company Collection.)

NOYO RIVER MILL of McPherson and Wetherby's Noyo River Lumber Company built in 1861 by Tichnor and Hendy. In 1883 Robert Byxbee acquired an interest and when Tichnor died in 1883, Byxbee merged his ownership with Joseph Clark to form Navarro Lumber Company. The mill burned in 1886 and was rebuilt farther up the river. The panic years of 1893 to 1895 ruined the business and the mill remained idle. It burned in 1902 after a spotted career of forty years and the holdings were sold to C. R. Johnson and associates. (Watkins photo from Bancroft Library, University of California.)

WHARF AT LITTLE RIVER with schooners *Point Arena*, left, and *Celia*, right. (Photos Escola Collection.)

LATE VIEW CASPAR LUMBER COMPANY plant. The enterprise had a long and stirring career, operating continuously from 1864 when Jacob G. Jackson bought it from Kelly and Randle until it shut down in 1946. Owners in 1880 were F. A. Wilkins, Henry Fisher, Charles G. Jackson and E. Sweet. In 1891 Mrs. Annie E. Krebs took over the management which later passed into the hands of her sons C. E. De Kamp and C. J. Wood. (Photo Union Lumber Company Collection.)

ARTISTIC SPLASH IN CASPAR CREEK—Camera catches pattern of erupting water as big redwood is chuted down. On log is pond man Clarence Freathy. (Photo Escola Collection.)

TWO KINDS OF DONKEYS—one a Dolbeer road engine using manila rope and purchase blocks—the other a docile water packer. (Photo Caterpillar Tractor Company from F. Hal Higgins Collection.)

CASPAR ROADING ENGINE
Compound cylinder logging engine used in the '20s, Federal truck on track was used as switching engine. (Agricultural Extension Collection, University of California.)

SIERRA NEVADA BIG TREES

"In the 1920s, when the General Grant Grove was about all that remained of the Kings River Big Trees," wrote Lizzie McGee in the March 1952 bulletin of the Tulare County Historical Society, "a young man came from Iowa to visit friends. They took him to the Grant Tree. With a spool of number eight thread he encircled the trunk and when he returned to the valley, he made it into a circle on the level yard. He shook his head and said: 'I was going to send this back to my mother and tell her I measured a tree and that this string is the exact circumference. But now I won't send it for she believes I am her truthful boy.'"

California's storied mountains, the Sierra Nevadas, have one distinction not always accorded them. They produced the biggest redwoods. This is the *sequoia gigantea*, a larger variety than the Coast redwood, *sequoia sempervirens*. The 8000 acres in the Kings River and Converse Basin areas were at one time dominated by these big red-barked beauties which grew to thirty feet in diameter at the base and three hundred feet high.

The first sight of these great bulks so impressed the California pioneers and imbued them with such fierce pride, they stripped the bark from a redwood

(Continued on Page 89)

KAWEAH COOPERATIVE COLONY in 1884 was a brave try at socialized living in the redwood forest but failed because of human frailties. Trials and tribulations beset it from the start and it officially gave up in 1891. Smith and Comstock operated a sawmill for the colony. (Photo Bancroft Library, University of California.)

PROTECTOR OF THE FORES

Jesse Hoskins was a dedicated man. A ranch near Tulare, he spent his summers in the Sierr cutting fence posts cut of the fallen sequoias ar came to regard himself as a shield against the e croaching menace of the sawmills. About 1897 filed a claim on 80 acres of choice timber to pr tect it from the axe.

One of the big trees in Jesse Hoskins preser had been burned to a shell but was still growir and this was to be his temporary home. He us a ship's auger to bore holes in the heart and ch eled out blocks of wood. From these he fas ioned miniatures which he sold to the summ campers swarming into the area. In the meantir he built a habitable room in the hollow of the tre

It was about twelve feet in diameter with sor seven feet of head room and it furnished shelt and certain sanctum to Jesse Hoskins. When finished the work on it, he christened the "Herc les" and expected it to last forever.

Perhaps it will. Harold G. Schutt — himse a man dedicated to sequoiana, made the above ph to in 1955 and reported the hollowing out proce had in no way affected the big tree's vitality. S drips from the walls and the room is always dar but this helps protect it and who can say th Jesse Hoskins' Hercules Tree will not remain f the last man to see?

HERCULES TREE—HOME OF JESSE HOSKINS who in 1897 dedicated himself as a shield against the encroachment of the sawmills. Standing by tree is Fred Wells, 101, veteran of early shingle mills. (Photo by Harold G. Schutt.)

REDWOOD FROM MOUNTAIN HOME—Twelve-horse teams fed lumber from mill to railroad. Jack Doty on left wheeler. Scene about 1905. (Photos Harold G. Schutt Collection.)

ROADING BIG LOG IN KINGS RIVER CANYON — Loggers anxiously watch action as log rolls into chute. (Photo by Harold G. Schutt Collection.)

in the Calaveras Grove in 1854, reassembled it on a frame and tried to prove to a skeptical world there were trees this big. But people said: "It can't be. It's just another boast of that brash young State of Californy." But the desire to impress the world with the redwoods' size continued strongly, matched only by the practical desire to cut them for lumber. Both activities prospered until there was little to show or cut.

The cutting began as early as 1856 when Smith and Hatch built a sawmill on the Whitaker Ranch in northern Tulare County at Miramonte. There were others in the Eshom Valley area in the '50's and '60's, with the R. E. Hyde mill in Whitaker Forest in 1870.

Orlena Barton Wright, writing in the Tulare County Historical Bulletin, said:

"My father came to Tulare County in the fall of 1865. Most of his summers after that were spent helping build and run saw mills, The Hyde Mill, now Whitaker Forest; the Wagy Mill at Meadow Flat. It was here at Meadow Flat that Bud Barton made the first Sequoia into lumber,

but here are his own words published in the Fresno Bee, November 28, 1926.

" 'In 1869, I, myself saw the the first Sequioa Gigantea ever made into lumber in the big tree belt.

" 'However, this tree was not felled by the woodman's axe. On New Year's night, 1868, this tree slipped its moorings in what is now known as Whitaker Forest and floated down Eshom Creek lodging a mile below Meadow Flat. For 2 or 3 weeks previous to New Year's day, it had been raining and the whole side of the mountain north of Eshom Creek had slid in an avalanche into the creek damming up the waters. On New Year's night the dam broke and the whole mass came down in a mighty rush. The trees and rocks may still be seen strewn all along the canyon of Eshom Creek, below Redwood Mountain.

" 'This tree was only six feet in diameter. If it had been much larger we could not have handled it with the milling equipment then used. At that time I was sawyer in the old Turbine Mill built by Jasper (Barley) Harrell. With a double circular saw we cut the tree into three logs lengthwise. Abe Murray, Sr. spoke for the lumber before the tree was hauled from the creek. Murray had the lumber hauled to Visalia and with it built his house on the Murray ranch in 1870.' "

In 1865 a water-power sawmill was operated by J. R. Hubbs in the Dillonwood area on the North

Tule River. N. P. Dillon purchased the mill and
converted it to steam. The lumber was brought
out on a tramway using wooden rails, cars hauled
up grade by mules, running down by gravity. Later
this system was replaced by a flume which terminated just north of the old control station where the
Balch Park Road starts up the mountain.

"The second mill to come into the Tule River country,"
says the historical bulletin, "was brought by ox-team from
Santa Clara county about 1870, by Charles F. Wilson.
It was set up at Happy Camp on the headwaters of
Rancherie Creek. J. Kincaid bought it at auction in 1876
for $400.00 and a year later sold it to Rand and Haugh-
ten. A. M. Coburn bought Rand's interest and operated
it for several years. It was moved several times, finally to
a point below Mountain Home. Coburn built a flume
down Bear Creek and his 'dump' on Haughten's place.
(Afterwards this was known at the Pete Planchon or
Jake Garner place.) For many years Coburn operated a
finishing mill at Springville, near the Soda Spring. Later
he was County Clerk of Tulare County.

"L. B. Frazier built the first mill at Mountain Home.
Frazier, a promoter of some ability, built a road from Milo
and Rancherie up Bear Creek to the site of the mill. He
moved a mill from the Pine Ridge area above Tollhouse
in Fresno County to Mountain Home and operated it a
short while in 1885. Charles Doty helped haul the mill
and drove bull teams to skid logs to the mill. Frazier went
broke and left the county. The mill passed to Pease, Newport and Jerrard. Frazier came back in 1889 and block-
aded the road he had built, hoping to collect tolls, but
some of those who had not been paid for their work tore
down the barricades and in the confusion the county ac-

quired title to the road. It is still used as a fire protectio[n]
road but is very steep. The Frazier mill burned in 188[8.]

"The Enterprise mill, the largest in this area, was erected in 1897 about a mile above Mountain Home, but operated only a short time because the company had onl[y]
80 acres of timber. The sawdust piles just above 'Hercules,' the tree with the room cut out of its heart, indicate the site of this mill.

"Charles Elster purchased the Coburn mill in 1898 an[d]
later purchased other mills and consolidated them jus[t]
north of the present buildings at Mountain Home. A M[.]
Conlee operated a mill at Brownie Meadow which is als[o]
in the immediate vicinity.

"These mills cut pine, red fir and redwood. Very fe[w]
big trees were cut that exceeded sixteen feet in diamete[r.]
They were cut ten or twelve feet above the ground, tw[o]
men ordinarily would fall a big tree in two or three day[s.]
An undercut was made, it was sawed from the opposit[e]
side and wedged over. The redwood being brittle was often badly shattered in falling and the logs had to be divided into sections either by splitting or blasting in orde[r]
to go through the saws in the mills. Most of these mill[s]
had two circular saws, one above the other and slightl[y]
behind. For a time the Dillonwood mill had a 'splitter,['
a long drag saw that cut big logs endwise.

"Lumber from the Mountain Home mills, except Coburn's, was hauled down to the valley by teams. Th[e]
mountain teamster was a skilled man and guided h[is]
animals by word of mouth as much as by his 'jerk line[.'
Lumber at the mill was about ten dollars per thousand.
Many of the old houses in the county are built of redwood from these mills and generally are still quite soun[d]
because termites do not attack this lumber.

"Many people used to spend their vacations around th[e]
mountain sawmills. Lumber to build a little cabin didn['t]

MILL AT HUME from rim of dam which created Hume Lake. (Below) Geared locomotive hauling short car of ties for extension of railroad running to Hume Mill. (Photo Harold G. Schutt Collection.)

cost much and the grain farmers, when the harvest was in, went to the hills for several weeks. Fred Wells, of Tulare, recalls that six or seven hundred people camped about Mountain Home and in 1887 three babies were born there. J. J. Doyle started his Summer Home resort in 1890 and actually sold lots for cabin sites. This area is now Balch Park.

"There was one mill on the South Fork of Tule River about two miles below Rogers Camp southwest of Camp Nelson. It was probably started in the 70's and in 1884 was being operated by Porter Putnam. Coburn is supposed to have purchased this mill."

In the '60's, Sequoia Lake was called Mill Flat Meadow into which Mill Flat Creek tumbled down from the General Grant Grove area. On the northern rim the People's Mill operated, machinery for which had been freighted by oxen and wagon from Visalia.

After the owners quarreled, Joe Hardin Thomas bought the mill at sheriff's sale and ran it from 1865 to 1867. He used bull teams for logging and cut several million feet of lumber which the oxen hauled to Visalia.

In 1881 Smith Comstock contracted with S. Sweet and Company of Visalia, who controlled the Wagy Mill, to cut lumber at Hitchcock Meadows near Happy Gap above Meadow Flat. In 1883 Comstock bought the mill and moved it to Big Stump near the present entrance to General Grant Park where he cut redwood and sugar pine,

probably fir and yellow pine. The mill was moved downstream several times and once was operated by J. C. Stansfield. Smith Comstock controlled mills in Mexico and Atwell's Mill on the Mineral King road. Comstock Lodge, owned by his daughter, Mrs. Effie Simmons, stands close to this same general area.

THE BIG TREES IN BIG TIME

Smith Comstock was making lumber of the redwoods but his operations seemed pitifully small to two plunging industrialists in San Francisco — Hiram C. Smith and A. D. Moore. In 1885, astounded and inspired to action by the thought of vast expanses of virgin redwood and sugar pine, they laid plans to attack the timber by big-scale methods.

They first bought the old Abbott Creek Mill and cut pine for a 54-mile waterway to be built overland down to Sanger. At the same time they started construction of a new sawmill on the higher level at Millwood. To insure a continued supply of trees they acquired large tracts of land in Converse Basin, Indian Basin and the Hume areas.

"To get this vast acreage," writes Lizzie McGee in 'Mills Of The Sequoias,' "they took advantage of an existing law that permitted a United States citizen to file on 160 acres of timbered government land. There were few requirements attached. You saw the land, you made a filing, and in due time could pay a flat sum and get your title. A few early settlers did this in good faith. But in this instance, most people considered an unfair advantage was being practiced. Agents scoured the nearby valley towns, looking for men who had not yet used their timber claim rights and who were willing to bargain for a few easy dollars. They gathered them up by the stage loads. As many as eight loaded vehicles were observed passing mountain homes in a single caravan. Arrangements were made for meals at farm houses. The men were taken to the vast beautiful timber belts where many of them had their first glimpse of the magnificent redwoods. They took a birdseye view of them — probably admired them. Large township maps were spread out, showing desireable quarter sections of vacant land. Each picked himself out a claim and they were taken back to Visalia where a U. S. Land Office was then located. A stop there, and with a scratch of a pen, they were in possession of a filing on a timber claim where some mighty big trees had flourished for some thousands of years. They had enjoyed a free mountain trip. It was presumed — and some said positively — all expenses paid. There were those who refused to enter into the deal who claimed they had been promised "everything free up to the possession of a deed." In course of time the acreage price to the Government was paid off, his deed was delivered and the trees were his. But not for long. There was a wholesale transfer of deeds to the Company,

for which each owner received a modest sum, in exchange for his timber right inheritance. All perfectly legal as far as the law was concerned, that is, if he didn't advertise his pre-arrangement concerning the deal. The Company was all set to begin slashing trees."

The logging itself was a fabulous enterprise in that day. The Converse Basin was in effect a great ampitheatre about six square miles in area, a thousand feet deep, out of which the logs had to be hauled. Here were great thousand and two thousand ton mammoths, weakened by axe, saw and wedge, brought to earth with a thundering, shattering thud, shafts so brittle the impact crushed and ruined fifty percent of the wood. Yet the onslaught went on steadily, bull teams replaced by steam donkeys, log chutes built to railside. A narrow-gauge railroad was built out of Converse Basin to upper Abbott Creek where a big power hoist raised the multi-tonned logs over the ridge and lowered them to the transfer station from which they were hauled to Millwood.

The logging camp at Converse, the mill at Millwood and the Abbott Creek mill, constituted a crew of over two thousand men in 1890, who worked about eight months a year when weather and snow conditions permitted. And the town of Millwood prospered. The McGee manuscript records:

"Visitors and campers swelled the population and Millwood continued to grow. It had a summer school of forty or more pupils, with Mrs. Poley Kanawyer as their teacher. There was a post office, store, butcher shop, a blacksmith, a doctor's residence, barns, cook-house, personnel quarters and numerous summer shacks for working men and their families. The Sequoia Hotel stood south of the town and above the road that went on up to the Lake. Bud Parker of Visalia came on and rented a large company building and converted it into the Red House Hotel — meals 25 cents, beds, 25 cents. The Sequoia Hotel changed its name to "The White House." A cow pasture and slaughter pen up the road a quarter of a mile supplied fresh beef for the army of people that came and went, and the working crews that made up old Millwood. A short cut on the Millwood Road branched off just below the slaughter pen and dropped down to the creek below the lumber yard to join the Lower Mill Road. This was known as "Butcher Cut." The inevitable red light district had its own little village off a mile or more to the south.

"The Smith and Moore headquarters stood up

ROLLING REDWOOD SECTION AT CONVERSE BASIN preparatory to splitting — in order to handle and go through saws. Note axes and auger hanging on log, latter used to bore powder holes as in view below. (Photos Harold G. Schutt Collection.)

the creek from the mill. Up there many Gay '90 parties made plenty of merry entertainment for the elect few and their friends. From cities came those who sought recreation in a mountain lumber camp city. Stories of gay living went the rounds.

"Along about 1899 the Smith and Moore Millwood mill was moved to Converse Basin. Most of the freighting and moving operations was done over the railroad and up the hoist. A makeshift road permitted some teaming over the top and down to Converse. This mill was a large concern. Here would be cut the really big trees. Reducing a ten, twenty or more foot sequoia log into sizeable proportions for transportation to the mill, and for running through the saw, required cruel blasting. With long augurs, holes were sunk deep into the heart of the log, the hole was packed with black powder or dynamite and a long fuse attached. Sometimes the explosion split the log more or less evenly but many more times it was wastefully shattered. The loss of valuable timber was pitiful. In time a splitting saw was invented and the loss was not so great. (Ed: This splitting saw was a band, the log resting on two carriages, one on each side of the saw.) Even so the woods were a sorry

looking shambles when cutting operations were completed. About as much waste timber covered the ground as was cut at the mill. Some claimed

even more. A tremendous cut was made at this mill. Magnificent groves of sequoias were depleted to extinction.

"As the lumber was cut it was loaded on flat cars, sent to the hoist by cable, and on down to the railroad, to be coupled on to an engine for delivery at Millwood where it was dried and then flumed to the Valley. In 1902 a small mill was built on Upper Abbott Creek where the cable track met the railroad engine. This mill cut the remaining pine trees in that area."

But it was the lumber flume to Sanger which staggers the imagination. The fifty-four mile viaduct ran from Millwood, down Mill Flat Creek to Kings River, down river, then crossing it by suspension bridge above Trimmer Springs, thence west to Sanger. The building of the great water course took three years and cost $300,000 — a crippling amount in the '80s except to men like Hiram Smith and A. D. Moore. Henry McGee, contractor of Orosi, superintendended the work and considering the rough territory over which the flume was built, the small loss of life during construction was remarkable. Sometimes the workmen were suspended from cliffs and many times they worked on trestles sixty to eighty-five high. One was a hundred and eight feet from the ground, skirting a boulder-bedded canyon five hundred feet down. The "Mills Of The Sequoias" paper describes the flume in fascinating detail.

"The flume was a veritable lumber ditch, with sloping sides and a flat bottom. With proper di-

mensions to carry a train of clamped lumber, or narrow boat or donkey, that conformed to th◼ shape of the flume, with not too much play, o◼ which a man or two could easily sit.

"The supply base with which to fill the flum◼ was the waters stored up behind the Sequoia Lak◼ dam. The flow ran rapidly coming down to Mil◼ wood, where a slack grade reduced it to a slo◼ moving gentle stream. In the lumber yard her◼ men were clamping lumber. It was arranged i◼ bundles about twelve by twelve inches square an◼ eighteen or twenty in length. Iron clamps ◼ somewhat large proportions were hooked over eac◼ end and wooden wedges were driven in to tighte◼ up the slack. This insured the bundle against fa◼ ing apart during the long rough trip down.

"Lumber over twelve inches in width, choi◼ sugar pine and redwood, was hauled to Sanger ◼ wagons. The greater width because the flum◼ could not carry it, the choice lumber because t◼ rough passage would injure its soft, fine texture.

"Like the launching of a ship, bundle aft◼ bundle slipped easily into the slowly runni◼ stream. Herders were on hand to hook them t◼ gether into trains of three bundles. One by o◼ the trains were released to begin their long journ◼ to Sanger. They moved smoothly over easy grad◼ went racing down cascading inclines, over eigh◼ five foot trestles, and along steep cliffs where t◼ flume was anchored to the overhanging rocks ◼ cable. The course followed was down the sou◼ side of Mill Flat Creek, past the lower mill, wh◼

LANDING AT SMITH COMSTOCK MILL—1886 with redwood log being side-hitched down slope. Photo courtesy of Effie Comstock Simmons, daughter of Smith Comstock. (Harold G. Schutt Collection.)

other bundles were hooked on to take their trip downward.

"Rancheria Flat, Camp Number Three, was the next levelling off stop. Here another train of three bundles were hooked on. Then down to Kings River they sailed, and reached Camp Number Four and One-Half at Cow Flat, where another train of three was hooked on.

"Nearing Maxon's Ranch (Trimmer Springs) the flume came down an elevated incline to the top of a high structure over a bridge that spanned the mighty Kings and crossed over to the north side. From here on the wooden ditch was set on high trestles that paralleled the river to several miles beyond Centerville where it turned westward into Sanger.

"At suitable points along the flume route, where slackening waters permitted, stations or camps were maintained. There were fifteen in all. From these stations flume walkers patrolled the ditch on wooden catwalks attached to, and a part of, the side of the flume. These trouble shooters were constantly on guard against lumber jams, breaks in the flume, or obstacles that might obstruct free passage of lumber trains. Their business was to see

that the lumber kept moving. Overlooking steep mountain cliffs, deep ravines and dangerous rocky canyons, these flume walkers made their way cautiously and by some were dubbed 'flume snakes.' They each carried a picaroon, a sort of lumber tamer.

"It is said that a man by the name of Bowell invented the picaroon idea but they were fashioned at the lumber company blacksmith shop from double-bitted axes. One bit was only slightly reshaped and the other hammered to a curving point. With this contrivance the flume walker reached out and straightened into line a train if it got wild and tried to buckle up and block the stream. As a lumber train reached the several stations along the way, it was halted and the next down-coming train was hooked on by the herders. By the time the valley floor was reached, great trains of lumber bundles moved gracefully and slowly into the Sanger lumber yard. They were slid off and piled up to dry, in readiness for finishing operations at the milling plant nearby. The flume had a capacity of 250,000 feet of lumber daily.

"A telephone service covered the entire fifty-four mile stretch. In case of accident, or flume

DILLONWOOD MILL—ABOUT 1904 — This was a small operation on the North Tule River. Small shed in center in front of mill housed log splitter shown on next page. (Photo Harold G. Schutt Collection.)

trouble, contact could be made with either terminal, and with stations along the way. Jimmy Mansfield of Centerville was superintendent of the flume crew.

"On steep inclines there was considerable slopping out of water. To replenish this waste, feeder streams were run in by small flumes. At Millwood "Little Lake" on Mill Flat Creek above the mill, furnished the first supply. Rancheria and Cow Flat each furnished a supply, with probably others.

"There seems to have been two types of boats. A long boat that formed to the sloping sides of the flume, and a dinkey, braced underneath in W shape, each section appears as a V-shaped culvert. The dinkey accommodated two men, front one with feet braced against a backstop, the second behind and braced against him. There were no side rails or front or back housing. You just braced yourself, grabbed hold the side of the teetering dinkey and took off. It must have been a breathtaking experience on some of the steep tumbling sections. These boats were used by company inspectors, the supply crew, and in emergency by

(Opposite) YARDING SPLIT SECTIONS of redwood in Smith Comstock timber—1888. Third man from left is Jesse Pattee who helped cut General Noble Tree for World's Fair display. Billy Megee on left was killed by rolling log. (Curtis photo from Harold G. Schutt Collection.)

others. For use between Millwood and Maxon's, supplies were freighted up from Sanger and shipped down by water. Clamps and boats were sent back by freight.

"A ride down the flume was a thrill to be remembered. At times it was dangerous. The load was balanced to dip the rear end slightly. Before starting a boat down, a poster was attached to a lumber clamp, signifying such an intention. No more lumber was sent down till the bundle passed all the stations and arrived at Maxon's. An "all clear" was sent to Millwood. The boat was released and away the passengers went on a wild, wild ride. At any of the camps they could, if they chose, crawl out and proceed on foot down the catwalk. It is not reported that one ever did so.

"Coming down the rapids, the boat with the speed of fifty miles per hour, outran the water. When it scraped the bottom the speed was lessened. But if the rear end kicked up a bit, when the onrushing waters overtook the boat, an awful spill could result. Such an incident happened on a steep stretch of rapids below the lower mill. Two men with two suitcases were in the midst of a flume 'shoot the chutes' joy ride. The boat was winning the race. With a burst of speed the water bumped it from the rear. The whole outfit, boat, men and suitcases took a terrific spill. The suit-

97

CHUTES INTO DILLONWOOD MILL
—(above) Four small flumes gathered water to form one large one to carry lumber down mountain. (Left) Log splitter was long saw driven by crank on countershaft. Note rope belt. (Photos Harold G. Schutt Collection.)

cases kept on riding the stream. By the time the passengers righted themselves and got underway their luggage had been fished out at Rancheria Station where the owners picked them up.

"There is the story of a Company doctor, in answer to an outside call, who took the thrilling ride, some of it after nightfall. A nurse reports having taken the ride with an expectant mother. She recalls it as the most exciting experience she ever went through. It took five or six hours to make the trip to Maxon's. Here passengers alighted and took a stage on to their destination. Better time could be made with a greater comfort, for from here on boats and lumber moved along at about four miles an hour, as the more level floor of the valley slowed them up."

"When the trees had all been slashed in the Converse Basin area, a track was laid to a saddle overlooking the upper Kings River Canyon. Here a hoist was installed. From there the track was laid down into Indian Basin (in the vicinity of the junction of Kings River Highway and the Hume Road). This was the Rob Roy Chute and was operated with a hoist. In a year or so the Converse Mill partly burned down."

The Smith and Moore era came sharply to an end. The mills were operating at full speed and capacity, cutting about one hundred thousand feet a day, and while they owned much more timber farther up Kings River, the capitalists seemed to consider it too expensive to bring it in. What they actually thought is not known but it is reasonably sure that in spite of the wealth of timber they had cut, they had done little more than to break even financially.

"Along about 1907," continues the McGee story, "George Hume came into the picture. His father was a successful lumber baron, operating on a large scale in the forests of Michigan. With some little knowledge concerning the first rudiments of the lumber industry, and with the backing of his father, George plunged into the business way out here in the beautiful Sierra Nevadas.

"George Hume came on when the destruction of the Converse Basin trees was past its peak. Even yet surroundings probably looked big to him. It could have been quite natural that he was drawn into a false sense of a great abundance, by the vastness of it all. Nor would it have been unusual, had he lapsed into the early Californian custom of relaxing and 'taking it easy.' It was generally claimed that pleasure before business ruled his activities out here in the west and that money from home kept his lumber industry going.

"He started out in a big way. With him was a Mr. Kessner. They negotiated with Smith and Moore to purchase their interests, land and mill, including the vast holdings over in the Ten Mile Creek area. A Mr. Bennett was hired to complete the deal and manage the setting up of a mill on the creek. In exchange for his services he took stock in the new company. The Smith and Moore Mills became the Hume-Bennett Lumber Company. Somewhere along here the industry was known for a time as the Kings River Lumber Company.

"A small mill was set up in a beautiful meadow over on Ten Mile Creek. With this, lumber was cut for the forms for the Hume Lake Dam and for lumber to construct a large new mill. The Converse Mill was moved over by freight wagons. To accomplish this a road was cut out over the hill and down into Indian Basin. From there it climbed over the ridge and down into Ten Mile Meadow. It wasn't much of a road but answered the purpose. A special wagon was made by Gil Braden of Centerville — equipped with four wheels in front, modern truck style. It probably was the first of its kind ever to be made. Twelve mules were required to haul the load. Josh Benedom of Sanger did the driving. Hume Lake Dam was finished in 1908, with a roadway along the top that led out into the timberlands to the east. Andy B r o w n superintended the job.

"In the meantime a makeshift road was run from the slaughter pen to M i l l w o o d Road, down Butcher Cut to the lower end of the lumber yard and on across Mill Flat Creek. It crossed the railroad and passed the cookhouse, and wove a steep, narrow path up the hillside to the north grove of sequoias, thence up through the park to join a skeleton road on out to Cherry Gap. From there this road followed the north hillside and connected up with the Indian Basin trail over the hill and down to Hume.

"Freight, machinery and railroad iron came by twelve and fourteen mule teams up the Millwood Road and on over the new trail. It took a good, well-trained team with a good driver and lots of 'Gee Bill' and 'Haw Jim' and much jumping the chain to maneuver a string of fourteen mules over the narrow roads and short turns. But animals and men knew how. But it was safest if small outfits listened to the bells and screeching of wagon metals and found a good turnout and waited till the big team passed by. In 1907 Bert Belknap of Reedley drove a fourteen mule team, hauling railroad iron up the difficult roads to Hume. The company put many miles of railroad out into the

START OF FAMOUS FLUME AT MILLWOOD — This Smith and Moore operation was sometimes called the Upper Mill. It cut about 90 thousand feet a day, the better grades hauled out by wagon, most of it flumed fifty-four miles down Mill Flat Creek to Kings River, thence to Sanger. See text. (Photo Harold G. Schutt Collection.)

timber northeast of Hume Lake over which to bring logs to the mill. By September, 1909, the Hume-Bennett Mill, modern in every respect, was in operation. The lake was filled and logging operations were going on in full force.

"A flume out of Hume Lake dropped down the steep, rocky hillside. It paralleled the rapids of Ten Mile Creek most of the way, on its race to the Kings River, thence down the river to join Mill Flat flume at the junction of the two streams. Here a small lake and a feeder flume replenished water wasted on the way down. This stretch of flume covered more wicked territory than the Sequoia Lake flume. It cost more money to build and had more and longer steep grades. From Hume to Sanger it covered seventy-three miles, the longest in the world, it was claimed. One trestle was a hundred twenty feet high and several more seventy and eighty-five. Stretches where the flume was anchored to rock cliffs were common. Over a million feet of lumber was required each year to repair the flume. Several workmen lost their lives by fall-

ing into the rocky ravines. One man fell eight[...] five feet into a boulder-strewn canyon. There w[...] nothing but a mangled roll of humanity to pick u[...]

"The stretch of flume to the river was so stee[...] it never was full of water. Only the most rec[...] less and daring ever attempted the ride, exce[...] where necessary. A boat left the water far behi[...] and looking back, the passenger saw a waterl[...] flume bottom. But woe to him if he bore down [...] the rear end of the boat too hard and in scrapi[...] the bottom tarried too long. Sliding along midw[...] between the waters that were racing on ahead [...] him out of reach and those that were tumbli[...] down from the rear to overtake, could be dang[...] ous. John Perry, constable at Hume, rode down [...] serve some legal papers somewhere on the li[...] Near Camp One and One-Half his boat upset a[...] he took a terrible beating. He lay in bed over th[...] weeks nursing cuts and bruises, and probably so[...] broken bones.

"Far down on the flume toward the river a l[...] ber jam caused a breakage in the flume. Bef[...]

SANGER FLUME WAS NEVER PHOTO-GRAPHED but where it ran along the ground it looked something like this. On most stretches it clung to high cliffs or bridged Kings River, water pushing "trains" of lumber at five or six miles an hour. It took three years to build it and cost $300,000.

shipment could be halted, a mountain of lumber piled up in a creek bed. The spot was inaccessible and salvaging was out of the question. Bennett proved unsatisfactory as a lumbering manager and withdrew from the firm. About this time, perhaps earlier, the concern was reorganized and became the Sanger Lumber Company.

"Prior to 1914 the state built a road from the General Grant Park north boundary, over Cherry Gap and on to Hume. It was an improvement on the first Hume makeshift road. It continued on around the point and into Hume as the present road now does, thus eliminating the steep climb up, over and down into Hume. This state road crossed over the Hume Lake Dam and continued eastward along the hillside for six miles. It ended

at a jumping off place that proved to be too great a barrier to consider further building. Out on the dead end, one is faced with deep chasms, high projecting granite ridges and rugged mountainsides. Densely covered uncountable acres of brush and small timber reaches out eastward and down, northward to Kings River. This state project was terminated as World War I loomed up ahead. Newly elected Governor Hiram Johnson came in for undue criticism when he refused to go along with more state money to finish the road into such forbidding territory, only to land far above and up the river, missing all the scenic beauty of the river. Fifty thousand dollars per mile had already been spent on the project with only a skeleton road to show for it. The war would have

TRAMWAY FROM CONVERSE BASIN MILL to Millwood where lumber was flumed to railroad at Sanger. (Photo Harold G. Schutt Collection.)

halted construction work anyway. In the meantime the present Kings Canyon Highway has been routed and was later built down near the beautiful Kings.

"Hume became a popular summer camping place. Boating, fishing and hunting with surrounding scenic beauty unsurpassed, drew many tourists to add to the company population and Hume became quite a camp village during the summer months. In 1917 the mill burned down and never was rebuilt. It is claimed that it took ninety-five thousand dollars of Dad's money to clear up the Hume debts. Buck McGee, who had been head blacksmith during the Hume milling operations, was retained as overseer, and as salesman for the personal property around the mill site.

"J. A. Brattin of Sanger had charge of the settling up of the Hume interests. The mountain timber lands and the vast logged over acreage was disposed of and became again the property of the U. S. Government. It came under the supervision of the Department of Agriculture, and within the boundaries of the Sequoia National Forest. Brattin says that when he took charge of the Hume business in Sanger there were many tons of earlier records, no longer needed, destroyed. Yes, the mills of the sequoias operated on a huge scale. Big trees were wasted and little, if any, profits made. Most companies quit in the red.

"In 1923 and again in 1928 forest fires swept up from the Kings River and burned many miles of the flume. In time the entire flume was dismantled. The mountain stretches were razed and burned to dispose of a fire hazard. Buck McGee

says he was the last man to take the thrilling ride down the Ten Mile stretch and on down to the Kings River. He took a speedy way to go fishing. He walked back. He says he is glad further temptation to repeat the ordeal is removed. Going down wasn't bad but the long, long walk back was awful."

BOOLE TREE FOR POSTERITY

It is said John Muir passed beneath the great canopy of a "majestic old scarred monument" in the Kings River forest. "Four feet from the ground" he wrote in ringing words, "the tree was thirty-eight feet eight inches in diameter inside the bark. I was not able to determine its exact age but undoubtedly the tree was in its prime, swaying in the Sierra winds, when Christ walked the earth."

Authorities concluded the tree Muir referred to was the Boole Tree, named by Dr. A. H. Sweeney of Fresno, in honor of a good friend, Frank A. Boole, superintendent of the Converse Mill operation. The loggers had undoubtedly passed up the monarch because its enormous weight, in falling, would have broken the fibres so thoroughly a great proportion of the wood would have been unfit for lumber. The Sierra Club, headed by Joseph La Conte, appealed to the government at Washington which authorized an exchange of timber for the Boole Tree which still stands as a silent monument to the past when nature alone ruled the wilds.

102

FALLING THE MARK TWAIN C. C. Curtis wrote on the back of this photo, "Beginning an eight-day job of cutting down the Mark Twain sequoia tree to ship to Central Park, New York, and London, England, Tulare County, California, 1891." The remarkable photograph on the next two pages shows the Mark Twain falling. Note falling bed of branches and side "undercut," necessary for limited saw lengths, even though two saws were brazed together. (C. C. Curtis photos Harold G. Schutt Collection.)

TOO BIG...OR TOO BEAUTIFUL?

Mark Twain Tree Goes Down in History

Nobody knew exactly why the big tree was left. In the early 1800s Smith Comstock was logging the Kings River canyon and while he had a sentimental interest in the giant sequoias, he was a lumberman with at least one eye on profits. There might have been another reason—that the logs would be too big to go through his sawmill—but he could have blasted them in half as other loggers did.

At any rate history records that Smith Comstock spared several big trees (which were later cut by others) among them the Mark Twain giant which had a special destiny. Comstock decreed that it be carefully felled and sections sent to world famous museums. He made arrangements with Collis P. Huntington to make formal gifts of them to the American Museum of Natural History in New York and the British Museum in London.

The exhibit sections measured sixteen and a half feet inside the bark. The balance of the tree was made into fence posts, the stump remaining as a memorial near the present main entrance to the Giant Grove in Kings Canyon in National Park.

103

VAN DOORMAN EXHIBIT—In 1892 Neal Van Doorman started a one-man project to ship sections of a big sequoia which he had felled to some Eastern city. He was a reticent man, mysteriously silent about the details but a printed source at the time stated that he planned to build a room out of 16 sections, some elaborately carved to accommodate 100 persons. The pieces were apparently shipped out but where erected is not known. Van Doorman is man standing. (Photo California Redwood Association.)

SECTIONS AND STUMP MARK TWAIN TREE with chips almost burying it. The section above was split into segments as shown on opposite page by using a dozen wedges. Note improvised saw—two ten-foot blades brazed to make one. (Curtis photos Harold G. Schutt Collection.)

DOWN A LUMBER SLUICE

(This article by H. J. Ramsden, concerning the adventures of famous San Francisco capitalists, appeared originally in the New York Tribune, is reprinted here from Wood and Iron, March, 1888.)

The flume is built wholly upon trestle work and stringers; there is not a cut in the whole distance, and the grade is so heavy there is very little danger of jam. The trestle work is very substantial, and is undoubtedly strong enough to support a narrow gauge railway. In one place it is seventy feet high. The highest point of the flume from the plain is 3,700 feet, and on an air line, from beginning to end, the distance is eight miles, the course taking up seven miles in twisting and turns, it being fifteen miles from beginning to end. The total fall from top to bottom end is between 1,200 and 2,000 feet, or an average of 120 feet to the mile. The sharpest fall is three feet in six. There are two reservoirs from which the flume is fed. The flume was built in ten weeks. It required 2,000,000 feet of lumber in its construction.

Upon my return I found that Mr. Flood and Mr. Fair had arranged for a ride in the flume, and I was challenged to go with them. Indeed, the proposition was put in the form of a challenge — they dared me to go. I thought that if men worth $25,000,000 or $30,000,000 apiece could risk their lives I could afford to risk mine, which was not worth half as much. So I accepted the challenge and two boats were ordered. These were nothing more than pig-troughs with one end knocked out. The boat is built, like the flume, V-shaped, and fits into the flume. It is composed of three pieces of wood—two two-inch planks sixteen feet long, and an end-board which is nailed about two and one-half feet across the top. The forward end of the boat was left open, the rear end closed with a board—against which was to come the current of water to propel us. Two narrow boards were placed in the boat for seats, and everything was made ready. Mr. Fair and myself were to go in the first boat, and Mr. Flood and Mr. Hereford in the other.

Mr. Fair thought we had better take a third man with us who knew something about the flume. There were probably fifty men from the mill standing in the vicinity waiting to see us off, and when it was proposed to take a third man, the question was asked if anybody was willing to go. Only one man, a red-faced carpenter, who takes more kindly to whiskey than to his bench, volunteered to go. Finally everything was arranged. Two or three stout men held the boat over the flume and told us to jump into it the minute it touched the water, and to "hang on to our hats." The signal of "all ready" was given, the boat was launched, and we jumped into it as best we could, which was not very well, and away we went like the wind. One man who helped to launch the boat, fell into it just as the water struck it, but he scampered out on to the trestle, and whether he was hurt or not we could not wait to see.

The grade of the flume at the mill is very heavy and the water rushes through it at railroad speed. The terrors of that ride can never be blotted from the memory of one of that party. To ride upon the cowcatcher of a locomotive down a steep grade is simply exhiliarating, for you know there is a wide track, regularly laid upon a firm founda-

FIFTY MEN ring Mark Twain stump. (Curtis photo Harold G. Schutt Collection.)

tion, that there are wheels grooved and fitted to the track, that there are trusty men at the brakes, and better than all, you know that the power that impels the train can be rendered powerless in an instant by the driver's light touch upon his lever. But a flume has no element of safety. In the first place the grade can not be regulated as it is upon a railroad; you can not go fast or slow at pleasure, you are wholly at the mercy of the water. You can not stop, you can not lessen your speed, you have nothing to hold on to; you have only to sit still, shut your eyes, say your prayers, take all the water that comes—filling your boat, wetting your feet, drenching you like a sponge through the surf—and wait for eternity. It is all there is left to hope for after you are launched in a flume-boat. I can not give the reader a better idea of a flume boat ride than to compare it to riding down an old-fashioned eave-trough at an angle of 45 degrees, hanging in mid air without support of roof or house, and thus shot a distance of fifteen miles.

At the start we went at the rate of about twenty miles an hour, which is a little less than the average speed of a railroad train. The reader can have no idea of the speed we made until he compares it to a railroad. The average time we made was thirty miles per hour—a mile in two minutes for the entire distance. This is greater than the average time of the railroads. The red-faced carpenter sat in front of our boat on the bottom, as best he could. Mr. Fair sat on a seat behind his, and I sat behind Mr. Fair in the stern, and was of great service to him in keeping the water, which broke over the end board, from his back. There was a great deal of water also shipped in the bows of the hog-trough and I know Mr. Fair's broad shoulders kept me from many a wetting in that memorable trip. At the heaviest grade the water came in so furiously in front that it was impossible to see where we were going, or what was ahead of us, but when the grade was light, and we were going at a three or four minute pace, the vision was very delightful, although it was terrible.

In this ride, which fails me to describe, I was perched up in a boat no wider than a chair, sometimes twenty feet high in the air, and with the ever-varying altitude of the flume, often seventy feet high. When the water would enable me to look ahead, I would see this trestle here and there for miles, so small and narrow, and apparently so fragile, that I could only compare it to a chalk mark, upon which, high in air, I was running at a rate unknown on railroads. One circumstance during the trip did more to show me the terrible rapidity with which we dashed through the flume than anything else. We had been rushing down at a pretty lively rate of speed, when the bow of the boat very suddenly struck something—a nail or a lodged stick of wood which ought not to have been there. What was the result? The red-faced carpenter was sent whirling into the flume ten feet ahead. Mr. Fair was precipated on his face, and I found a soft lodgment on his back. It seemed to me that in a second's time Mr. Fair, a powerful man, had the carpenter by the scuff of the neck, and had pulled him into the boat. I did not know that at this time Fair had his fingers crushed between the boat and the flume.

But we sped along; minutes seemed hours. It seemed an hour before we reached the worst place in the flume, and yet Hereford tells me it was less than ten minutes. The flume at this point must have been very near 45 degrees inclination. In looking out before we reached it, I thought the only way to get to the bottom was to fall. How our boat kept in the track is more than I know. The wind, the steamboat, the railroad train never went so fast. I have been where the wind blew at the rate of eighty miles an hour, and yet my breath was not taken away. In the flume, in the bad places, it seemed as if I would suffocate. In this particular bad place I allude to, my desire was to form some judgement of the speed we were making. If the truth be spoken, I was really scared almost out of reason; but if I was on the way to eternity, I wanted to know exactly how fast I went, so I huddled close to Fair and turned my eyes toward the hills. Every object I placed my eyes on was gone before I could clearly see what it was. I felt that I did not weigh a hundred pounds, although I knew in the sharpness of the intellect which one has at such a moment, that the scales turned at two hundred.

Mr. Flood and Mr. Hereford, although they started several minutes later than we, were close upon us. They were not so heavily loaded, and they had the full sweep of the water; while we had it rather at second hand. Their boat finally struck ours with a terrible crash. Mr. Flood was thrown upon his face and the water flowed over him, leaving not a dry thread upon him. What became of Hereford I do not know except that when we reached the terminus he was as wet as any of us.

This only remains to be said. We made the entire distance in less time than a railroad train would ordinarily make it, and a portion of the time we went faster than any railroad train ever went. Fair said we went at least a mile a minute. Flood said we went at the rate of 100 miles an hour, and my deliberate belief is that we went at a rate that annihilated time and space. We were a wet lot when we reached the terminus of the flume. Flood said he would not make the trip again for the whole Consolidated Virginia mine. Fair said he should never again place himself on an equality with timber and wood, and Hereford said he was sorry he ever built the flume. As for myself, I told the millionaire that I had accepted my last challenge. When we left our boats we were more dead

than alive. The next day neither Flood nor Fair were able to leave their beds, while I had only strength enough left to say, "I have had enough of flumes."

BIG TREE CAMERAMAN

With skill and great perserverance, C. C. Curtis made a lasting record of the Tulare County redwoods and lumbering activity, although his efforts met with many frustrations. His story comes to light through his daughter, Mrs. Cecile Pennell, and the Tulare County Historical Society.

He came from Marshalltown, Iowa, b e f o r e 1880 and set up a "picture gallery" in Hanford— Curtis and Tandy — in 1884. He became a member of the Kaweah Colony and worked on the project road to the mountains. After the colony break up, he was one of the committee of five to investigate the possibilities of colonizing the Kettleman Plains.

Curtis photographed the Big Trees by making frequent forays into the timber, carrying equipment and sometimes wife and daughter Cecile on donkeys. He spent eleven years in and out of the mountains, doing other photographic work in Traver, Porterville, Esperanza, Big S t u m p and Millwood. The full recording of his General Noble sequence is shown in this volume.

His health failing and camera work unprofitable, Curtis took up work in a San Jose spice mill and when he left in 1899 he buried his glass negatives under the floor. He later worked in the San Francisco plant of J. A. Folger and was one of the four men who saved the mill from burning during the earthquake. He then moved to Berkeley, subsequently to Pasadena. The Curtis family had expanded, members gathering around Cottage Grove, Oregon, and they wanted their father to come with them. Before he went north he had word the old spice mill in San Jose was to be demolished and he hurriedly retrieved the old negatives from the cache under the floor. Many of these have since been lost but those used here came from the collection of grandson C. C. Annand of Bremerton, Washington.

The camera used for Curtis' photographs in the timber weighed forty-five pounds with six plate holders. 8x10 glass plates were used with time exposures. Negatives were developed at night in a tent with the aid of a red lantern and prints were made on albumin paper sensitized by a silver nitrate solution floated on just before use.

GENERAL NOBLE GOES TO THE FAIR

Of the several efforts made by California pioneers and enthusiasts to put the size and story of the giant sequoias before the world, by far the most ambitious was the display prepared for the World's Columbian Exposition at Chicago in 1893. This became a consuming project, five months in actual cutting and shipping, and the steps were officially photographed by C. C. Curtis.

"The tree selected by the committee," writes Harold G. Schutt in the Tulare County Historical Bulletin after an interview with Jesse Pattee, one of the choppers and participants in the whole program, "was the General Noble tree, growing about three miles north of the northwest corner of General Grant Park.

"The tree was cut off about fifty feet above the ground and Jesse Pattee will never forget the experience of the moments when the tree fell. It did not go as planned. The trunk slipped back onto the stump and broke the scaffold. Pattee and the three others doing the cutting jumped onto the stump but couldn't stand even in the middle for twenty minutes because of the vibration.

"The tree was hollowed out and fourteen foot 'staves' cut with the bark and about six inches of wood. Then a section about two feet thick was cut off the remaining stump and then another set of fourteen foot staves was prepared. These parts were all marked and crated and hauled out to Monson for shipment to Chicago. The stump, fifty feet above the ground, was nineteen feet six inches larger at the top of the twenty foot stump which was left when the job was done. This relic is now called the 'Chicago' stump.

"Efforts have been made to find out what happened to the exhibit after the fair closed as it was understood that it had been moved. After hearing that it went to Washington, Senator Knowland was asked to secure such information as he could. The following was furnished by the Legislative Reference Service of the Library of Congress under date of March 16, 1950.

"The section of Giant Sequoia originally cut for exhibit at the World's Columbian Exposition at Chicago in 1893, and later set up on the grounds of the Department of Agriculture in Washington, D. C., was dismantled and moved to the Arlington Experiment Farm on the Virginia side of the Potomac River during the winter of 1931-32. It was never re-assembled, but rested there in storage for several years. No record of its final disposal is available, but it seems to have been destroyed. No one was found who could show whether this was before or after January 30, 1942, when the Army took over the Arlington Experiment Farm. Prior to this date, all the agricultural work with the records of research had been moved to the Agricultural Research Center at Beltsville, Maryland.

"The record is clear that the section was moved to the Arlington Experiment Farm, but thereafter nothing concerning disposal of the section could be found. Moreover, no piece of the section was placed with the wood exhibit in the Smithsonian Institution.

"The following statement is copied from page 32 of the Official Record, United States Department of Agriculture, for January 30, 1932:

"The 30-foot high section of the trunk of one of the giant California Sequoia trees which has stood on a concrete base in The Mall in Washington, in front of the Department of Agriculture main building, for the past 38 years, has been taken down and stored at Arlington Experiment Farm, just across the Potomac from Washington, to keep it out of the weather until another suitable location is found for it. This huge hollow cylinder of timber, which may possibly have been a husky sapling in King Solomon's day, 10 centuries B. C., was moved because it hindered the Government building program.

GENERAL NOBLE RECREATED at World's Columbian Exposition at Chicago, 1893. This 30-foot hollow cylinder built from segments cut from General Noble Tree, as pictured on these pages. After the exposition it was placed on the Mall in Washington, D. C., later removed to storage at Arlington Experiment Farm.

"The section was cut for display in the Government exhibit at the World's Columbian Exposition in Chicago in 1893. The tree grew on the boundary line between Fresno and Tulare Counties, Calif. From records which Dr. W. A. Taylor, chief of the Bureau of Plant Industry, was able to find, the tree stood some 300 feet high in the forest. The section was hollowed out in California, and the great hollow pieces of wood was cut for shipment to Chicago, each piece being equivalent to a log 4 to 5 feet through and 14 feet long. The section was cut 30 feet above the ground level where it grew, and it is 26 feet in diameter and 85 feet in circumference. (Note: Pattee recollection is 20 feet which checks with stump). At the exposition the section was set up in the Government Exhibit, a spiral stairway was run from the bottom to the top on the interior, and many visitors to the fair viewed the exhibit from the vantage point thus afforded.

"After the exposition the section was shipped to Washington, and in 1894 it was placed in The Mall where it has been an object of interest to the thousands of visitors to the National Capital who come its way."

Continues Harold Schutt, "It is hoped that the exhibit has not been destroyed. Probably the bark was falling from the trunk after 38 years in the open in Washington. The picture seems to indicate wires around the tree to keep loose bark in place. But this could be replaced at relatively small expense and some museum could have a marvelous 'Big Tree' exhibit. Incidentally the picture of the tree in Washington was found in some old files of the Tulare County Board of Trade, stored in the basement of the Visalia Municipal Auditorium."

C. C. CURTIS official photographer of the General Noble project, was not paid for his participation. He expected to sell prints at the Fair, spent many months and dollars to accompany the exhibit but Fair authorities would not permit sale of pictures. He suffered such a financial loss he was forced into other fields to make a living.

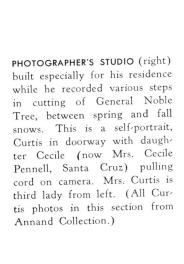

PHOTOGRAPHER'S STUDIO (right) built especially for his residence while he recorded various steps in cutting of General Noble Tree, between spring and fall snows. This is a self-portrait, Curtis in doorway with daughter Cecile (now Mrs. Cecile Pennell, Santa Cruz) pulling cord on camera. Mrs. Curtis is third lady from left. (All Curtis photos in this section from Annand Collection.)

MAKING UNDERCUT (above) of General Noble, forty-five feet from ground. (Top right, center) Top of tree split as it fell. Part of tall stump has been hollowed out, platform lowered to cut 30-foot section in half. (Below, right) Platform lowered again, segments ripped, one shown as it is being lowered to ground.

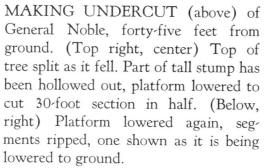

CLOSE UP OF GENERAL NOBLE at stage shown opposite bottom right, debris piled on platform. Bear was shot by one of party. (Below) Numbering segments and loading on wagons for trip to railroad.

CREW ARRIVES WITH PACK TRAIN to prepare parts of General Noble for shipment by wagons and train to Chicago. (Below) Eight mule-and-horse teams, each bearing one casket-like segment of big tree. Smith and Moore, who donated timber and paid labor costs, received no financial compensation and little acclaim.

REDWOOD AND THE CITY

San Francisco, says the legend on the old chart, was built on sand and mud and the ashes of six great fires. It was built by the labor of Irish, Chinese, Russians, Finns, Mexicans and Portugese. It was built with the profits of hides, gold, lumber and grain and general foreign trade. And with pine and stone, it was built and rebuilt of redwood, both as a building material and as an industrial commodity.

On any day in the '80s the coasters were steaming south from the Humboldt and Mendocino shipping points, deck-high with redwood, some anchored in San Francisco Bay with pay loads going over the side to lighters. In their company were four-masted schooners and three-masted b a r k s waiting to ship redwood piling and timbers and set sail for Australia and the Sandwich Islands.

Some lumber cargoes barged over from Sonoma mills and those scattered back of beyond, were being discharged directly on the wharves and shoremen were piling the redwood on wagons. Then horse teams would haul the loads up Nob and Telegraph Hills, imperiously preserved by the mansions of Hopkins, Stanford, Huntington and Crocker, now being encroached upon by prosperous middle-class solidity. Most of these new houses were big, over-stuffed in the Victorian manner, but to the world on San Francisco Bay, their wooden neo-gothic scrollwork and five-story frills spoke loudly of money in the bank.

The whole bay was alive in the morning. A tug was nosing the *William H. Smith* into her berth and two others were bringing in the big Panama mail ship. Salvage boats, cattle carriers, dredges

GLITTER ALLEY — A San Francisco showplace of the '60s—Gilbert's Melodeon at Clay and Kearney (Portsmouth Square). All photos in this chapter from W. Stanley Collection, Bancroft Library, University of California.

COBWEB PALACE AT MEIGG'S WHARF owned and operated by Abe Warner in the 1880s. Warner reads newspaper in the company of two monkeys and three parrots. He was a colorful showman tending toward the bizarre.

and garbage ships moved across the belt of fog drifting down from the inland rivers. Side-wheelers stern-wheelers and scows crossed the path of the ferry *Solano* which was loaded with freight cars, Port Costa to Benecia. On the Oakland side the Pacific Coast whaling fleet rode at anchor near the decrepit remains of old clipper ships.

The tip of the San Francisco peninsula was saw-toothed with piers and the great sea wall, begun in 1873 and built of the rock blasted out of the hills, had now become the two-hundred-foot wide Embarcadero. Warehousemen by the thousands were sorting, checking, piling merchandise a n d material. Sailors, engineers, bargemen and pilots drift along the busy waterfront. Masters, cooks and stewards weave in and out, buying galley stuff for long voyages to Chile and Sumatra.

Teams rattle over the planks and shouts of the drivers mingle with strident notes of steam whistles and the solid banging of cargoes dumped on the docks. Newsboys shout raucously and carters hawk their wares into the unheeding ears of Kan-akas, Chilenos and Spice Island ship hands. Hundreds of sounds and hundreds of smells greet the cosmopolitan workers of the city up from Montgomery Street.

Loggers, fishermen, miners and sailors pick up the sounds and blend them with the babble of foreign tongues and the strains of the hurdy gurdy in the deadfalls between Stockton, Washington and Broadway. This is the Barbary Coast—grease trap and catch basin for those long restrained in the redwoods, on ships and in shafts.

The sounds now are the clinks of bottles and chips, the wheedling cries of pimps and prostitutes, the cackling sing song of Chinese peddlers, the shrill squeaks of flutes and fiddles. The smells are those of incense, fish and musty drink emporiums —perhaps the oily, rich smell of opium cooks.

Where there are loggers there are women and in the saloon are red-faced, frowsy ones, sad eyed ones from Spain, hoarse-voiced strumpets f r o m who-cares-where. This is "The Roaring Gimlet" with bar on one side, sofas across the room and a curtained archway at the back. In others — "The Bull Run", "Cock Of The Walk", "Star Of The Union' — are more women, including long-dressed Mexicans and silk gowned Chinese in their blue

SOMETHING OUT OF POE or De Quincy was the interior of Abe Warner's Cobweb Palace. Loggers, sailors and world drifters drank to the screech of cockatoos and eyed the cannibal curios.

satin shoes, black braids tied under their chins. They were bought as children in Canton for $40, sold in San Francisco for $400 and never knew any other life.

The signs say "Dance", "Dixie Club", "Green Wave Music Hall" and "Dugan's Drinks And Dance." And more women and more whiskey — all kinds, all ages, all prices. Behind a "lodging" placard, a wrinkled old harridan sits in a cubicle and collects four-bits for a night's hideout, leading the way with an oil lamp down a creaky-floored corridor to a room filled with bunks like a sailing ship's fo'castle.

Small gas lights shine dimly over the long tables in the gambling rooms which take another chunk of the logger's money. Two hundred men crowd the poker, chuck-a-luck and faro tables. And no lights at all in the alleys, fringed with flesh and little wicketed windows. Vagabonds, thieves and rum heads stand and wander vacantly — some of them loggers who forgot the woods and can not find their way back to fresh air.

WHAT CHEER HOUSE

"After a winter on Feather River," wrote Idwal Jones in Westways for May, 1957, "with icy floods, chilblains and gravel of diminishing returns to contend with, young Bob Woodward bade farewell to its scenery with no twinge of regret. He left, working his way down in the skiff of a friendly hunter, who was conveying a load of elk and black bear to the market in San Francisco. Woodward had a large acquaintance with hunters. That he was to make use of later. He was a Massachusetts youth, strictly brought up, abstemious and industrious. These may seem rather chill merits, but in him they were combined with tact, politeness and attention to personal advantage.

"A few days after his arrival in the city, with pounds of gold dust, he visited a dock, where sacks of coffee were being auctioned. He bought a number of them, and on hearing that the lease of a small shop in cobble-stoned Waverly Place was on sale, he also bought that, and set up business with a coffee-house. He waited at the tables in apron, he carried off the dishes, made change, was often the echoing voice from the kitchen, booming until long after midnight. At daybreak he was at the market buying elk, ducks and sturgeon, all at favorable rates.

"His coffee-house, prosperous from the start, had to be enlarged twice, but was still as plain and work-a-day as a butcher's block. Then no more could it be enlarged.

LOGS CRACKLED AND HOSPITALITY PREVAILED in Bob Woodward's famous What Cheer coffee house on Sacramento Street below Montgomery, in San Francisco in '60s. Inn had no bar but excellent food and friendliness and made its owner a fortune.

Congested with high Tong buildings, temples, balconies convenient for flying kites and lowering strings of firecrackers; and lined with barber-shops, each with the symbol of the craft before the door, a tin basin on a chair, Waverly Place had also taken on another name, Fifteen Cent Street. The cost of a Chinese haircut, including the brushing and rebraiding of the customer's queue, was 15 cents.

"As Woodward pondered a shift elsewhere, a convenient fire destroyed half the city. Few cities can match the example of San Francisco, which had been burned down so often since the gold rush. The flames of one pyre were visible as far away as Monterey, 80 miles distant. Though the fires caused much annoyance, the rebuilding of the city was in the nature of an improvement.

"A chart of that yet-smoldering period bears the legend: 'San Francisco was built on Sand and Mud and the Ashes of Six Great Fires. In 1853 over 300 Brick and Stone Buildings Arose.' On Sacramento Street, just below Montgomery, was Woodward's new house yclept What Cheer House. It was of brick from around the Horn and stone from China; five stories high, with a florid ironwork balcony, and before it a row of columned gas-lamps. Productive of wonder was its old-established look as though prospering there for a hundred years.

"Logs of huge girth crackled in the lobby fireplace. What Cheer House was the first hotel in California to serve dinners *a la carte*. Patrons dined well at the charge of 50 cents; they could dine in baronial fashion at $15, but without champagne. What Cheer had neither a wine-celler nor bar. Expectant diners had, anyway, their accustomed bars to visit for a nip or two enroute. A party of them, including two judges, had a narrow squeak one night. "Careless shooting. Bullet grazes three gentlemen at the bar of the Hotel de Ville," went a headline next morning. The culprit, who had thought he was in a court of law, was released after an admonishment to behave more respectively.

"The lack of a bar seemed to have no effect on the hotel's revenues. The one drawback was that, since journalists and editors found that their duties led them elsewhere, the What Cheer got small reclame in the press. It had powerful rivals in the Tehama House and the Oriental Hotel, which had bars of magnificence, and there were bars more sumptuous, with no hotel attached.

"In 1854 one of the guests at the What Cheer was an ex-officer who had debarked from the Eureka boat, its sole passenger except for a crate of hens. Capt. Ulysses S. Grant had resigned from the army at fog-bound Fort Humboldt, in a spell of discouragement over his career. He was waiting until he could scrape enough funds to

SAILS FROM THE SEVEN SEAS in San Francisco harbor about 1900. Small tug in left center is the *Lottie*.

REDWOOD CASTLE on windy hilltop. Everything in sight of the Arthur M. Ebbett home, except the lawn, was built of California's own redwood. House was 1875 vintage on Jones, between Washington and Jackson.

MANSIONS OF THE MIGHTY were richly imposing and helped out the redwood mills. (Left) Frescoed and filigreed home on Polk Street. (Center) Mark Hopkins home, on site of present hotel, ran to turrets while Leland Stanford edifice, southwest corner of California and Powell (right) had columns and wide cornices.

return to Missouri, where he was to start life again in his father's leather-shop. For the time being, the attractions of the What Cheer, which were many, suited both his taste and his flattened purse. His room was moderately priced, and at the foot-stands in the hall he polished his own cavalry boots. Well-shod was well-enough dressed. When visiting Parker's Bank Exchange, the most famous of all the bars, he passed unnoticed in a Mexican cloak and worn slouch hat. Some of the army figures there like Halleck, Sherman, "Fighting Joe" Hooker, and a dozen other future generals, he knew by sight. He was to meet them again as commander of Lincoln's forces in the Civil War.

"On the second floor of the What Cheer was its library, the largest collection of books in the region. It had the novels of Scott, Dickens and Marryat; the classics, shelves of poetry, a hundred works on agriculture, and as many on military compaigns. More pretentious was the museum in the two rooms adjoining. It housed stuffed deer, eagles, condors, flamingos, marsh fowl, fauna of every sort on the coast. Also walrus tusks, Indian skulls, war clubs and pottery. The bulk of the specimens was gathered by Woodward's hunter friends. Dr. Friedrich Gruber, his taxidermist, was skilled also in the blowing and tinting of glass eyes.

"Here one heard l e c t u r e s on fish and sea-shells by Gruber, and on occasion (always a social event) a lecture by Prof. George Davidson, the astronomer and director of the Geodetic Survey. These talks, instructive in a high degree, were heavily attended by the residents at the hotel. "Happily," it was observed by a writer, "no bar nor gambling table, those instruments of vice and dissipation, were there to detract from the more respectable pleasures of the cabinet, the tastefully stocked bookshelves, the unrivalled collection of beach shells, and the Preparations Display, with whip-snakes, toads and four-legged chickens in jars full of alcohol."

"With his third fortune now made, Mr. Woodward, finding the role of boniface a shade insipid, turned elsewhere. He built and operated a tramway. He built a costly home in the Mission district, with a large deer park and a lake. A party he gave there for a troop of Civil War veterans with their wives and children, gave him the notion of turning it into a wonderland for the young. Terraces, a botanical garden, a pavillion for chariot races and skating, large enough to hold 5,000 people, sprang up. Also a zoo, a flock of mechanical swans with seats to guide over the lake, swings, merry-go-rounds, a concert stage, acrobatic shows, band concerts and nightly fireworks. The What Cheer's museum was brought over; an art gallery was built, with Virgil Williams in charge: a painter notable in his day, and the organizer of the city's School of Design. He taught Fanny Osbourne, later the wife of Robert Louis Stevenson, who dedicated Silverado Squatters to his friend, Virgil.

"Woodward's Gardens, a Vauxhall for the elders, and a paradise for the young, lasted for 30 years, when its proprietor retired to his estate at Oak Knoll in Napa Valley. The What Cheer House, itself built on the ashes of one fire, vanished in that of 1906.

"But there are mementoes of it at the Sutro Baths — some of Dr. Gruber's stuffed birds, a few Japanese fans, and the collection of sea-shells."

SOCIALISM IN THE REDWOODS

Out of a Knights of Labor picnic in San Francisco in 1884 grew the Kaweah Cooperative Colony, an experiment in socialized group living. The forest inspired and gave it impetus, human frailties brought it to a dead end.

Leaders of the project were Burnette G. Haskell, James J. Martin and John Redstone. They saw the timber as the economic means of sustaining a socialistic movement and organized an association of working men. Its purpose was to establish an industry which would support members living by the tenets of Christian socialism, insure them against want, provide comfortable homes and maintain the principles of justice and fraternity.

After several attempts to purchase land, they chose an area, close to the Grant Forest, in Kaweah Canyon in the mountains of eastern Tulare County. Fifty-three applicants filed but when they returned to pay, the government refused to accept the money in the fear that some monopolistic agency was behind the acquisition of the timber. Many of the aspiring colonists would not wait for the order to be removed, which they were sure

would be, and made "squatter" homestead claims, built a road and sawmill on monies raised by general subscription.

The first permanent settlers arrived at Kaweah early in 1887. There were seven camps of tents, the first at Arcady or Haskell's Bluff and after the bridge across the river was completed, at East Branch, Sheep Creek Flat, Advance, Camp Progress, Lookout and Saddle. When the road was finished the town of Kaweah was established.

Membership in the colony required the payment of $500. The initial cost was $10 and after contributing $100, a man and his family could take up residence and employment. The rest of the fee could be worked out in labor or paid with machinery or supplies. There were never more than four hundred active associates. Each settler had 150 square feet for his home, an eight-hour maximum working day for which he got 30 cents an hour minimum wage and no man received more than the next regardless of the kind of work.

The colony had the asset of Haskell's printing press and the weekly "Kaweah Commonwealth" was circulated thinly over the United States and England. There was postal service, library, school-

PORTER AND FLOOD MENAGES—In background is the stone mansion of mining magnate Flood, now the Pacific Union Club. The redwood home is that of the Porter family in the '80's on the present site of the Fairmont Hotel.

house, dancehall and meeting place, the blacksmith, carpentry and cobbler shop.

From the first the sawmill failed to cut to capacity and because of other troubles the colonists realized they must put it on a paying basis. Meanwhile the government had brought the trustees to trial for cutting timber illegally, fined them nominal sums but was unable to prove any cutting of sequoias. Nevertheless the public spread the word that the "queer people" were despoiling the Giant Forest.

A legal land site was secured from the heirs of the nearby Atwell estate, the sawmill rebuilt on the new ground, yellow and sugar pine cut to sup-

ply it. The U. S. Fourth Calvary appeared several times, finally stopping operations until the Secretary of Interior at length determined the mill was on patented land. But human inefficiency and wilfull violations of colony regulations defeated the mill and in July, 1891, it was an admitted failure.

By January of the next year the membership was so depleted the colony was officially dissolved. Founder Martin made several attempts to reorganize, recover damages and get a clear ruling from the government but none of these efforts was successful. The colonists drifted into other, more conventional fields and Kaweah was only a name of a story for grandchildren.

STOCK ITEMS FOR RICH SAN FRANCISCANS included redwood porch columns, newel posts and rails. (Photo Bancroft Library, University of California.)

HUMBOLDT HEYDAY

They were saying in San Francisco that the redwoods on the Mad River and the Eel River and elsewhere back of Eureka had the best chance to success. "They" were "big Eastern lumbermen" and they were backing up their claims by buying a lot of land. Especially they said: "This valuable tree here reaches the greatest perfection and attains gigantic proportions. Some specimens a r e found rising two hundred feet without a limb and having a diameter of over twenty feet."

Further, said these opportunists: "There are an estimated 500,000 acres of these trees and with the lumbering industry in its strong and vigorous youth, an almost unlimited field is here offered. One acre of Humboldt County redwood will yield as much lumber as forty acres of the average pine land."

The Eastern lumbermen were not alone in this judgement. They were joined by sea captains with business sense and California pioneers who came to farm and stayed to cut lumber. Hans H e n r y Buhne, William Carson, John Dolbeer, J a m e s

Ryan, James Duff, John Vance were a few of those who saw the redwoods as a power and profit potential.

The future of the redwoods had been forecast by the recorded visits of the Spanish missionaries and by the L. K. Wood account of Dr. Josiah Gregg's party of explorers in 1849. They were awed by the mammoth trees and while unable to do more than stare at them, believed they would someday form the basis of an industry as big in proportion as they were.

Two practical men of action who wanted to see this lumber industry get started built a small sawmill in Eureka in 1850 and gave it the Indian name "Taupoos" which became popularized to Papoose. They were Jim Eddy and Martin White and the mill was between the ends of J and K Streets. It lasted only a year as did the Luffelholtz mill and the 1852 enterprise of Byron, Deming and Marsh on the Trinidad River.

Another sawmill, built that year, was to have a more lasting significance — Ryan and Duff's

LOGGING CAMP AT ELK, Humboldt County in 1895.

A. W. ERICSON

FLAT OF PEELED REDWOOD ready for burning of slash after which logs would be skidded to rivers to pile up for spring freshets. (Vansant photo California Redwood Association.)

Eureka Mill. James Talbot Ryan, from Ireland and Boston, had originally led the Mendocino Exploration Company to Humboldt Bay and now joined with James Duff, fom St. Johns, New Brunswick, in an ambitious plan. He went to San Francisco, bought machinery and equipment, loaded it on the steamer *Santa Clara* and headed north with John Vance as mate and sailing master. While crossing the Humboldt Bay, the deck load shifted and went over the side. The loss all but killed the enterprise since the power plant of the mill-to-be was part of the lost cargo.

Ryan struck a desperation course. While he held the *Santa Clara* out in the bay, Duff, on shore,

ordered workmen to dig a channel (b e t w e e n Eureka's present D and E Streets) and then under a full head of steam, the ship rammed her way ashore. At once she was stripped of her steam engines which went into the sawmill and in six weeks the plant was turning out lumber from timber rafted into a tidewater slough, the bigger logs quartered for the circular saw. All this while the forty men of the crew, among them A. W. Torry, Frank Duff, George Carson and Alexander Gilmore, lived on the beachbound *Santa Clara*. More bad luck came the way of the Eureka Mill when two of the first lumber shipments were lost on the brig *Clifford* and bark *Cornwallis,* both foundering at sea.

In 1854 there were nine sawmills operating on Humboldt Bay. The Medina mill at Bucksport had also used ship's engines for power, those of steamer

GIANT SLICES FOR SHINGLE BOLTS —(Opposite) Bucking crew cutting and splitting redwood monarch. (Ericson photo California Redwood Association.)

125

HAMMOND MILL - SAMOA—Upper jack works carried logs to two band saws, one cutting redwood, the other Douglas fir. Smaller logs on lower slip went to double cut band at rear of mill.

Chesapeake — converted by the ship's m a s t e r Capt. Hasty Kingsburg Dodge. The Picayune mill was in Eureka, built by Hinkle, May and Flanders. The latter withdrew forming a partnership with Capt. Ridgeway in another sawmill project which became the Vance mill when sold in 1856 to Vance and Garwood. The Mulay Mill was formed by a company consisting of W. H. Dwyer, Thomas Graham, W. H. Sleymist, A. C. Moore and James Olmstead. The business passed through several hands, finally into the solid ones of William Carson.

The Jones, Little sawmill had its beginnings in the J. C. Smiley and Hiram Bean mill at the foot of Eureka's H. Street. It also had a succession of owners, the last being Jones a n d Kentfield. In 1853, Martin W h i t e who had accumulated $20,000, built the Bay Mill on Front Street between L and M. Three years later he had financial trouble and the mill was sold to John Dolbeer, who had come from New Hampshire and worked in the Salmon River mines, Daniel Pickard, Isaac Upton and C. W. Long. Later Charles McLean of San Francisco bought out the latter three partners and operated the mill with Dolbeer. When fire gutted it in 1860, the men rebuilt it, and when McLean drowned on the *Merrimac* off Humboldt Bar, Dolbeer joined with William Carson in this bay Mill No. 2.

Carson was a transplanted n a t i v e of New

LOG TRAIN AT BIG RIVER—Engineer on the head engine o f this Mendocino Lumber Company train was Andrew Escola. (Photo Escola Collection.)

DOLBEER DONKEY IN ACTION with three lines on redwood log, rolling it on car. Mule was water carrier. (Photo Worden Collection, Wells Fargo Bank History Room.)

HUMBOLDT SKIDROAD about 1910. Logs are moving toward donkey in background, haul back line shown at right center. (Photo Hammond-California Lumber Company Collection.)

Brunswick. He had worked in the gold mines and with Oliver Gilmore, Daniel Morrison, Sandy Buchanan and Jerry Whitmore, built the "Arkansas Dam" in the Trinity River. He was one of the first to recognize the merchantable value of redwood. He leased the Mulay Mill in 1856 and cut the first cargo from the big trees, shipping it out on the brigs *Quoddy Belle* and *Tigris*. The next year he joined with Phillip Hinkle in a sawmilling activity which proved unsuccessful. However, the 1863 partnership with John Dolbeer brought in a winner and resulted in a long Carson dynasty of successful operations and fame to the Dolbeer name in the invention of the steam donkey.

Over the many years, Dolbeer and Carson owned four sawmills and as each succeeded the other, only one feature remained the same — the mill whistle. It was the Carson decree that the original whistle always remain. The company owned a fleet of lumber schooners of which the most famous, because she became a "movie star", was the *Lottie Carson*.

An intimate view of Eureka in 1866 is given by John Carr in his "Pioneer Days." He mentions the sawmills and then goes into detail. "2nd Street was but little built upon. On the corner of F and 2nd, the Masons and Odd Fellows had built a two-story building, the largest then in town. Both of these orders occupied the upper story for lodge rooms. Ben Fiegenbaum was on the lower story with a general merchandise store. There were two smaller buildings between the corners of F and G Streets. That corner was occupied by an old building used as a boarding house for the John Vance mill hands. It was pulled down by Mr. Vance

when he built the Vance House. From the corner of 2nd and G, on both sides of the streets, were principally dwelling places as far as the old court house, the Humboldt Times on the south-west corner of F and 2nd. It was owned and edited by J. E. Wyman. On the corner of E and 2nd. C. W. Long and A. T. Gilbert were keeping a l i v e r y stable. John T. Young had a saloon on the opposite corner. R. M. Williams was running a l i v e r y stable on the corner of 2nd and D. What was known as Duff's Boarding House was on the corner of 2nd and C and was recently pulled down to make room for the Grand Hotel.

"2nd Street was then the main entrance to town. 3rd and 4th below E was not yet open, the ground a quagmire. So was E between 3rd and 5th. Rev. W. L. Jones and Major Long had the first buildings south of 6th on E. They had built a plank pathway over the marsh to get to their homes. All below 4th on D and C was partly brush and timber down to 2nd. 3rd Street was not yet graded. Huge stumps and logs filled the streets as far as 9th and there scattering trees and logs were seen until you reached the woods. In the upper part of town, all south of 4th was forest. Clark's Addition was not yet laid out and that part of town was forest, the woods coming down to 6th Street on D.

"Eureka has one steamer making two trips a month to San Francisco and several sailing vessels were engaged in the lumber trade and sometimes carrying freight and passengers. R. W. Brett's saloon, or "Brett's Court" as it was generally called, was general headquarters for all the wags in town. Every night one could hear all the

HORSES SECOND BEST TO BULLS but these logs in Newbury Mill Company's forest were small enough for an 8-horse team. Bulls had more sheer straining power and stamina, took abuse better. (Photo Pacific Lumber Company Col.)

scandals of the town retailed, news of the day discussed, cases in court tried, all manner of jobs concocted. It was the headquarters for all the sea captains and a jolly set they were, spinning yarns and having a good time generally. There were four other saloons in town at that time. There was one line of stages running between Hydesville and Eureka owned by Ballard and Swasey, making daily trips. And one line running between Eureka and Arcata."

Twenty years later the magazine Wood and Iron gave an account of a "redwood king's hospitality" when a party of pilgrims from San Francisco took a week's trip to the Humboldt.

"On Thursday, May 22nd, the steamer *Corona,* Capt. Hanna, made its way up to the dock at Eureka. There was nothing unusual about this; but still is was notable, because on the deck of that steamer was seen the portly form and the bearded face of one of San Francisco's most prominent lumbermen—a man known to almost everybody in the State, and one who seldom leaves the cares of business behind him. But there stood A. Powell, General Manager of the Puget Sound Lumber Company, casting his eyes for the first time in seventeen years upon the shores from whence come the millions of feet of redwood, so esteemed wherever known. As the steamer reached

the dock it became evident that his arrival was expected. Nearly all of Eureka had gathered to witness the arrival of the distinguished visitor, and on his landing he received a royal welcome. J. J. Loggie, on whose special invitation Mr. Powell made his visit, received the gentleman, and he was quickly escorted to the Vance House, where elegant quarters had been provided for him; in fact, he was as the bride led to the altar, for the bridal chamber had been especially fitted up for him. Like the loyal citizen that he is, he first paid obeisance to the powers that be, and paid a visit to his Honor, John Vance, the mayor of the city, who tendering him the hospitality of the little city, cordially invited him to attend the next Saturday a grand picnic, to be given by the lovely school marms and handsome children of Eureka. Surely the lines of the visitor had fallen in pleasant places. On Friday the party visited the Occidental Mill, rode through the vast forests on Ryan's Slough, then took guns and went in search of either chipmunks or deer, and brought up at the Freese Ranch, where good Mrs. Farrar spread a fine repast for tired and hungry mortals. After a pleasant drive the party found themselves in the evening at Vance's Hotel, where Isaac Minor, N. H. Falk, J. A. Sinclair, A. A. Curtis, D. J. Flanigan, David Evans, Josiah Bell and other noted lumbermen of Humboldt, gathered to make pleasant the visit of the stranger within their gates.

"On Saturday morning, armed as the law directs, fully equipped for the fray, the visitors made their way to

EEL RIVER AND EUREKA RAILROAD (above) running to Field's Landing wharf, 1884. This was the period when Pacific Lumber Company was owned by Paxton, Curtins, B. Law and James Rigby and was logging heavily. (Below) Pacific's mill at Forestville before 1895 fire.

Campton Park, a lovely spot on the Eureka and Eel River Railroad, where the school marms and their proteges were to picnic. Of course it was embarrassing for this simple-minded citizen to be at once launched into a sea of so much loveliness, for Eureka's school marms are noted for their many charms; but he stood the test nobly, and came out of the battle unscratched; this was owing, perhaps, to the fact that he had ever before him the staid example of Cousins, Loggie, and other reliable lumbermen, who have often met and been in the presence of the same guileless beauties.

"On Sunday, of course, the party attended divine service; that must never be omitted, and 'tis a peculiarity of Pacific Coast lumbermen well worthy of emulation that no matter where they are, if 'tis a possible thing, they attend church at least once each Sunday. On this day

1895 FIRE DESTROYED MILL AND LUMBER—When rebuilt, Forestville became Scotia and in 1898 Charles Nelson became president. (Below) Scotia general manager Ed Yoder, a legend in redwood lumbering. (All four photos Pacific Lumber Company Collection.)

Mr. Powell met an old time acquaintance, Simeon Zane, who in former years, way back in the sixties, resided across the bay at Vallejo.

"On Monday the magnificent country residence of Hon. John Vance was visited, where the party was warmly received by Mr. James Vance, the nephew of his Honor, the Mayor, who not only devoted his time to entertaining the pilgrims but provided them with the necessary accoutrements for fishing and a large quantity of splended trout was secured.

"Here on the summit of the hills, overlooking the valley of Mad River, Mr. Vance has erected a palace fit for the home of a king. Here, surrounded with works of art, not surpassed however by the works of nature, life can be enjoyed and perhaps true happiness realized. This mansion is presided over by a young lady, talented, refined, and highly accomplished, an artist, who amid the noble redwood forests is seeking that inspiration which alone comes to the born artist. Here a delightful day and evening was passed, and the next day saw the pilgrims departing for Eureka, passing through Arcata. One of the pleasantest events of the trip was the meeting of Capt. Harry Kingston and Mr. Powell, who years ago were schoolmates in Philadelphia, and who had not met for years.

"Tuesday night the fish caught on the trip were distributed among the hungry Eurekans, whose palate are not often tickled by such fine specimens of the finny tribe.

"On Wednesday another picnic was enjoyed by the party at North Fork, where again was temptation thrown in the path of our modern Joseph, but who once more escaped to tell the tale of venture by sea and flood.

"Jackson's well known mill was visited, as was Korbel

Bros., and Mr. Bower and Mr. Jackson showed them many courtesies and favors, but the charms of the picnic were too much, and after short visits with Lawyer Buck, Mr. Kirk, and Mr. Eberding, both Powell and Loggie were found dancing the stately minuet with North Fork's fattest and fairest damsels, until the shrill whistle of the locomotive warned them that a moment's delay would

see them forever lost or left behind. On returning to Eureka a pleasant visit was had with Capt. M. A. Brandt, of the schooner Occidental, whose generous lady kept the party supplied with flowers during their stay.

(Above) LOGS ON FRESHWATER SKIDROAD (Ericson photo) and below Scotia baseball team of 1902—most of names illegible. (Photos Pacific Lumber Company Collection.)

"On Thursday morning Dolbeer & Carson's bay mill was visited, when Mr. William Carson received them, and conducted them through their plant, after which the Pacific Lumber Company's mill was visited, President A. A. Curtis doing the honors, going so far as to improvise an observation car for their especial benefit. Here they visited the mill that turns out a 100,000 feet of redwood per day, and a shingle mill that produces 150,000 shingles per day. At Fried's Landing they were met by Mr. Henry Servier, of the Occidental Mill, with a team, and driven to Eureka.

"Friday was spent in fishing in Elk River, the stream that supplies the city with water, and fished from Showers' farm to the Brown claim, and returning prepared to bid farewell to the lovely little city where they had been so hospitably received.

"Saturday morning they bade farewell to all, and as the *Corona* steamed out of the harbor Capt. Hanna did his best to console the two weary pilgrims, who in a week's time had seen more, learned more. enjoyed more, and been more hospitably received than ever before in their long and eventful lives.

"They returned refreshed, reinvigorated, and the better able to take up the burden of life, fully impressed with the belief that not one half has been told of the beauties of redwood forests, and fully imbued with the idea that the future greatness of Eureka and Humboldt County is assured, and her generous, noble-minded citizens worthy of every joy and happiness vouchsafed to mortals."

MOVING 90 TONS IN UPRIGHT TREE at Pacific's Freshwater Landing No. 1. To get spar tree in proper place, it was topped 216 feet from ground, rigged, then sawed off at butt and moved by means of rigging and donkey. (Photo Agriculture Extension, University of California.) Right, center, a Pacific crew in 1905— extreme right front, Dan Angst, manager of Field's Landing from 1910 to 1944. Lower, top and bottom saws in Scotia mill. (Photos Pacific Lumber Company Collection.)

In 1865 Capt. H. H. Buhne joined with Capt. Nelson, Jones and Kentfield, of the old Smiley Mill, to form D. R. Jones and Company and build in Gunther Island the largest mill on Humboldt Bay at the time. Hans Henry Buhne was a dynamic character who had been a whaling master and come to San Francisco on the old *Clementine* to try his hand at placer mining. He fell desperately sick but finally recovered and took a second mate's berth on the schooner *Laura Virginia* bound for the Trinity River. On this trip he discovered the mouth of the Eel and was the first American seaman to enter Humboldt Bay. He remained to earn a living as a harbor pilot, built a big house on Buhne Point and became a merchant, running ships to San Francisco for goods. A shipwreck at Bodega ruined these efforts and after another period of sickness he became a professional deer and elk hunter, again going to sea as captain of the brig *Colorado*. At length he came ashore to operate

STEAM BUCKING SAW at chute of Standard Lumber Company in 1923. (Photo Agriculture Extension, University of California.)

steam tugs in the harbor and enter the lumber business.

By 1881 there were twenty-two sawmills cutting redwood in the county including the Occidental; Excelsior Redwood; Big Bonanza; Flanagan, Brosnan Company; Elk River Mill and Lumber Company; Eel River Valley Lumber Company; Isaac Minor's Dolly Varden; Jolly Giant; Jones Creek Mill; Trinidad Mill Company; Chandler, Henderson and Company. Korbel Brothers came on the scene in 1883, Glendale Mill in 1885, Northern Redwood Lumber Company in 1902 and after pioneer Isaac Minor retired in 1903, his children organized the Minor Mill and Lumber Company.

The Pacific Lumber Company at Forestville had its roots in the old 1869 operation of McPherson and Wetherby of Mendocino's Albion Mill. There was no logging done until 1882 when Paxton and Curtins of Austin, Nevada, with B. Low and

James Rigby of San Francisco, purchased it. The railroad extension was built to the Eel River and Eureka Railroad and lumber moved to Field's Landing, where a large wharf was built. In 1895 the mill was destroyed by fire, a new one built, the settlement name changed to Scotia and in 1898 Charles Nelson was elected president.

A major reorganization occurred in 1902. The Minnesota capitalist, H. C. Smith, acquired the controlling stock and that of the Excelsior Redwood Company, which became the Freshwater Lumber Company, and the former the Pacific Lumber Corporation of New Jersey. But two years later Smith resigned and E. M. Eddy of the Freshwater enterprise became head, that company discontinued. Again Pacific dropped logging operations until 1938.

The Hammond Lumber Company grew out of the John Vance Mill and Lumber Company at the hands of A. B. Hammond, another New Bruns-

LOGS PUNCHED OFF CAR at Goodyear Redwood Company, Greenwood, in 1923. Steam hose from locomotive was coupled to piston mounted on heavily weighted car. Placed behind logs the plunger rammed logs off car into water. (Photo Agriculture Extension, University of California.)

STARTING UNDERCUT on one of Isaac Minor's redwoods (Dolly Varden mill) out of Eureka. Even in this 1885 period Humboldt loggers were learning to keep cut close to ground if tree girth permitted use of crosscut saws. (Ericson photo California Redwood Association Collection.)

NEWBURG COOKHOUSE IN 1890—E. J. Stewart cut his eyeteeth at this Eel River Valley Lumber Company camp. Man is Robert Allen with Stewart's dog, woman and small girl, Mrs. Herman Doe, wife of logging boss and daughter Ruby. Other women—Angie Norman, Mrs. Jesse James, Mrs. Chamberlain, Clara Doe and Ruby Cousins. (Photo E. J. Stewart Collection.)

wick product who became a partner in a mercantile business in Missoula, Montana.

In 1890 when the Vance mill at Samoa and all properties including Eureka and Klamath River Railroad were up for sale. Hammond with five associates made the purchase. The firm was operated as the Vance Redwood Lumber Company until 1912, then changed to Hammond Lumber Company.

REDWOOD RECORDER

By trade a printer, by environment a logger and lumber worker, August W. Ericson developed his avocation into a business and became California's best know photographer of the redwods and Indians. He took thousands of views for commercial firms, individuals and for resale, processing them with painstaking skill.

In 1870 Ericson worked in the woods for the Trinidad Mill Company, then owned by the Hooper brothers, helped build wagon roads and skidroads in the timber, loaded cars with redwood lumber and ties and worked on the logging railroad. He informed himself of the country by taking stage trips to Arcata and Eureka and then quit the woods to clerk in a general store. All during these early years he carried on his printing trade in the off hours.

In 1876 he moved to Arcata to open a stationery-drug store under the name of Davis and Ericson. He augmented his income with the printing press and studied photography until he became adept in the use of the wet plate. When his brother Richard joined him, the two opened a printing shop and photographic studio. In 1888 he married a Eureka girl.

August Ericson was not a camera artist in the strict sense but was a diligent, careful workman who made faithful reproductions of the big trees and the processes by which they were converted into lumber products. His photos of the various Indian tribes people in ritualistic dances and village life rank with the best in the nation.

HALF CENTURY WITH REDWOOD

"I guess I saw all the boom days in the red-woods," says E. J. Stewart who spent fifty-two years with Dolbeer and Carson. "There was color and heartbreak and comedy that could happen only in these woods and the mill when tough Blue Noses came smacking up against tough Finns. I'm home-grown Eureka with red sawdust in my blood and I wouldn't have missed the show for all the tea in China."

E. J. Stewart did not miss much. He was orphaned at ten and went right to work. Herman Doe, logging boss of the Eel River Valley Logging Company, took him under one sturdy arm and his wife, who ran the cookhouse where seventy-five men stormed in three times a day, extended her protective wing. Sometimes Doe let him tend spool and sometimes drive the horse and buggy to Fortuna. That was 1890. Before Stewart was through with redwood, he had spent a lifetime tallying lumber, selling it in the yard, as shipping manager, assistant mill superintendent, logging manager and then general superintendent of the whole Dolbeer and Carson operation.

"I remember most of it," he says, "the sweating, steaming bulls dragging the logs to the landing in great clouds of choking dust, the water slinger scooping up big cans of water from the tubs along the skidroad and sloshing it in front of the teams. On steep grades, if it was muddy, he'd grab a shovel and throw dirt instead. Nine yoke of oxen were standard then. The Eel River Valley ran two side — each had a little donkey and riggers in a gulch and they were always racing to make their loads first.

"Falling was all hand work, choppers working in pairs. They built the falling stage ten to twenty feet up, above the swelling churn butt, and it took a day, maybe two, to fall a big one. They always cut the leaning trees first if possible so they wouldn't break if others fell into them, making a layout or bed for the heavy weight to fall on. They also tried to fall as many trees as possible uphill so they wouldn't be in the way of those going down.

"Ringers chopped circles around the fallen trees and the peelers came to strip the bark off. After the first fall rain the woods were set on fire, the slash burning up but leaving the redwood logs alone. That way is was easier to skid the logs out but the men were as black as coal miners.

RED SAWDUST IN HIS VEINS—E. J. Stewart spent 52 years in red-wood lumbering, retiring as Dolbeer and Carson general superintendent. (Below) Key men of Dolbeer and Carson organization—right to left, E. J. Stewart, logging boss Watt Hibler, Joe Meyers, sales director Henry Hink, Milt Johns and Lew Goddard. (Photos E. J. Stewart Collection.)

"After bucking the log fixers sniped the fore ends of the logs and smoothed the riding side for the skid road and chutes. Mighty skillful they were too. In steep country, the head swamper with a crew of ten to fifteen made roadways and chutes into which the little donkeys rolled the logs, the biggest one at the head. They'd throw chunks of wood in the chute as buffers between the logs, which were bridled together with hooks and rings connecting them. The bull team took hold of the

137

LUMBER SCHOONERS moored at Dolbeer and Carson wharf in Eureka. Cupola of Carson home at right behind trees. Below is original Dolbeer and Carson sawmill.

lead log and hauled the string to the cars. There was always a little grade provided on those chutes and roads as it was impossible for bulls to pull those heavy loads uphill.

"At that time logs were brought to the mill by a little gypsy locomotive that ran down and dumped the logs in the pond. It had a saddleback water tank on the boilers and at the back an overhanging platform for fuel wood. On the front end there was a gypsy rig which could be connected to a side spool, for loading and unloading logs.

"The old Newburg mill was a curiosity — double circular saw, ratchet and pinion drive, the carriage pulled back and forth by a cogged gear. The power was steam, from planing mill slabs, sawdust and shavings. The lumber went by railroad to Field's Landing and by boat to San Fran-

cisco. That was the day of sailing schooners, loaded by hand, piece by piece, without a chute.

"I went to Eureka after I finished school in '98 and started handling lumber for Dolbeer and Carson — and stayed there fifty-two years, retiring as general superintendent. It was a rewarding experience in every sense to be associated with those fine people — the Carsons, William senior and the three sons, John Dolbeer and others in the San Francisco office.

"John Dolbeer was a genius. He saw how slow and inadequate bull team logging was and invented the single cylinder, side spool donkey which was first manufactured by S. F. Pine Foundry and Machine Shop. It generated tremendous power and revolutionized logging methods on the Pacific Coast — one of the outstanding mechanical marvels of the industry."

DOLBEER AND CARSON MILL AND WHARF — photo taken from cupola of Carson home, shown in detail next page. (Photos both pages E. J. Stewart Collection.)

EUREKA LUMBER DOCK—1908. (below) Horses were the only motive power, work of tallying and loading heavy, the hours long. (Photo A. Quarnheim Collection.)

BIG REDWOOD butt log shows two undercuts were necessary to operate saw. (Ericson photo Calif. Redwood Assn.)

REDWOOD PERIOD PIECE—Elaborate and elegant was the old Carson home, built entirely of redwood in 1884. A famous Eureka tourist attraction, home is now a private club. (Andrews photo.)

FOUR JACKS—THE GIANT KILLERS — Dolbeer and Carsons choppers left to right—unidentified, Nick Sobol, Steve Rumi, Ed Murray. (Photo E. J. Stewart Collection.)

REDWOOD HAD ITS DETRACTORS

"This redwood lumber has some very valuable properties with others of opposed character. It is almost as brittle as glass and a 2 inch plank of it will not support the weight of one ordinary man. It splits with the least blow and is so soft I have known a small terrier dog, shut up in a new barn built of it, to gnaw a hole through the side and make his escape in a half hour.

"Some half dozen years ago a curious illustration of the unreliability of redwood appeared in San Francisco. Workmen engaged in putting a new asphaltum roof upon the three-story brick block on the southwest corner of Montgomery and California, and a drayman who had brought them some material, stood on a battlement wall looking

at them. Something attracted h i s attention, he stepped back and to the horror of the spectators, cleaved the wall entirely and fell in a perfectly upright position the whole height of the building to the sidewalk below. The crowd rushed in to see the mangled remains of the unfortunate man spread like a pancake over the sidewalk but to their astonishment,saw only a round hole in the planking about the size of an ordinary flour barrel. Looking down through the opening into the cellar which extended out under the sidewalk, they saw him pick himself up, walk to the stairs under the building and in a moment more emerge as sound and well as ever, not a bone broken or even a severe contusion received. The explanation of this remarkable occurance was simple. A part of the sidewalk was tough and hard Oregon pine but two or

three planks were redwood and he had s t r u c k squarely on his feet in the center of one, going through it like a 480 pound shot through the roof of a house.

"The county jail of Redwood City was built wholly of this brittle and thoroughly unreliable wood. As a matter of course, a prisoner who could command an ordinary table knife never tarried long within its walls! One night four or five prisoners who had been there some weeks left in disgust and the writer chronicled their escape for a San Francisco paper, stating incidentally that it was understood they dug their way out with a tablespoon and a tenpenny nail."

—from "California: Studies Of Life In The Golden State" by Col. Albert S. Evans, 1873

DOLBEER AND CARSON BULL ROAD near Fieldbrook using spools for cables on inside curves. (Photos Agriculture Extension, University of California.)

HANDMADE SUSPENSION BRIDGE over Yager Creek, suspended from a single, discarded logging cable. On property of Holmes-Eureka Lumber Company, Carlotta. (Both photos Agriculture Extension, University of Calif.)

1892 INVENTION OF DAVE EVANS, with E. H. Percy and Bethune Perry. They used snatch blocks on curves of skid-road to keep ground line under control. Both photos here show logs dogged together, being roaded to landing by donkey—on Excelsior Redwood operation. (Ericson photos Pacific Lumber Company Collection.)

MILL AT ANDERSONIA IN 1921 fifteen years after completion during which time it never cut a board, Anderson was killed in the building of the mill and 5 million feet of prime redwood logs lay in scummy pond. Machinery and saws were removed. (Photo Agriculture Extension, University of California.)

TWO WAYS TO HAUL LUMBER (Above) Three-wheeled traction engine hauling seven large wagons in Sierra Nevadas (Photo Calif. Redwood Assn.) and below, lone horse pulling car at dockside. (Photo Bancroft Library, U. of Cal.)

RAILROADS IN THE REDWOODS

In the gradual climb from Shake City you made spectacular curves, looping and doubling back in a variety of twists as the train crossed and recrossed the Noyo River. Some of the curves were so tight the brakeman had to unhook the safety chains between the cars to keep them on the track. You went up to about 1750 feet through redwoods flanking deep canyons and at Summit began descending more gently into the Willits Valley.

This was the steel that brought redwood logs out at Fort Bragg and hauled passengers and freight between that port and Willits to connect with the Northwestern Pacific Railroad, "the San Francisco train". The first passenger run was made on December 19, 1911 but the road had a background of logging experience dating from 1885 when owned by C. R. Johnson's Fort Bragg Redwood Company. It was only forty-two miles long but by all odds the most ambitious, colorful and longest lived railroad in the redwoods.

As early as 1855 the redwood interests started thinking about railroads and built in Arcata a mile long pier. The rails were 6x16 redwood timbers, at first topped with pepperwood 2x4's, then later by strap iron. On these a single horse drew a flat car loaded with passengers and freight out to berthed ocean vessels. This was the Union Plank Walk and Railroad Company and called the "Annie and Mary." As passenger business increased two locomotives were built for the run, one with an upright boiler, the other larger, with horizontal boiler and two oscillating pistons, this named the "Black Diamond."

A few years later John Vance, the Eureka lumber pioneer, tried to get right-of-way from that city up the Eel River Valley but failed. in 1873 he built a private line from his Big Bonanza sawmill to the Mad River slough. The railroad ran about five miles from tidewater along the north bank of the Mad, crossed it twice on 120 foot steel spans,

SONOMA ANTIQUE— Handmade engine and short trucks on Sonoma County logging operation in the '70s. (Cherry photo California Redwood Association.)

145

to the Lindsay Creek mill site, later named Vance and still later Essex.

This line was known variously as the Humboldt Bay and Trinidad Railroad, Humboldt and Mad River Railroad and more familiarly as Vance's Railroad. At the tidewater terminus lumber was loaded on lighters which were floated on outgoing tides to ships in deep water. By 1881 it boasted of a Baldwin locomotive and thirty truck cars, by 1886 had been extended four miles up Lindsay Creek.

When John Vance built a new mill on a point of land jutting out from the west shore of Humboldt Bay, named Samoa, he also built a railroad to serve it. It was opened to the public in 1897 and was given the name Eureka and Klamath River Railroad, running fifteen miles from Samoa through Arcata, to Vance and Lindsay Creek. By 1902, when the Hammond Lumber Company purchased the Vance interests, it had been extended through Fieldbrook to Camp 9, a distance of twenty-three miles. In 1904 the road was reorganized and renamed Oregon and Eureka Railroad, eventually reaching Luffenholtz and Trinidad with a branch from Little River Junction into redwood timber. A passenger could buy a ticket from Eureka to Trinidad and travel by the steamer *Antelope* to Samoa and then by train to the northern point.

The South Bay Railroad was an 1875 enter-

prise of Calvin Page, Charles Nelson and John Kentfield with other directors, H. H. Buhne and J. W. Henderson. The road ran from South Bay marsh up Salmon Creek. When the franchise to go up the Eel River was denied, the equipment was moved to Freshwater Creek and the line extended from this point to Humboldt Bay.

A narrow guage road was needed to penetrate the timber on the north fork of the Mad River and in 1881 the Arcata and Mad River Railroad began operating, hauling logs and passengers through Arcata. Directors of the road were G. W. B. Yocum, R. M. Fernald, B. Deming and E. A. Deming, and Austin Wiley.

When Korbel Brothers purchased the timber on the north fork of the Mad, the railroad was included in the transaction, the line extended to their mill and improvements made to the wharf. Several people were killed on the road in 1896 when a passenger train crashed through the bridge over Warren Creek. In 1902 the "Annie and Mary" was sold to the Northern Redwood Lumber Co.

Dolbeer and Carson built two miles of track in 1882 from Jacoby Creek to tidewater on which loaded flatcars rolled by gravity to the mill and were hauled back up grade by horses. Another short railroad here was the Trinidad Mill Company's. Originally built by Smith and Daugherty in 1871, it ran four miles from mills to wharf, was sold to California Redwood Company in 1883.

The Eel River and Eureka Railroad was organized in 1882 and three years later had completed a line from Eureka to Burnells. Two months later the Pacific Lumber Company completed a line

BECAME MUSEUM PIECE—California Western Railroad Car No. 43 carried many notables and lesser lights between Willits and Fort Bragg, was finally wheeled off to the Railroad Historical Society in Los Angeles. (Photo Union Lumber Company Col.)

from Alton to Scotia which began operation in April, 1896. This was extended to Elinor two years later and to Camp Nine in 1902, extensions

HAMMOND LOGS ROLLING INTO SAMOA—Schooners are moored at shipyard in left background, sawmill beyond. (Photo A. Quarnheim.)

EARLY HUMBOLDT LOGGING EN-
GINE probably about 1905 (Pho-
to Pacific Lumber Company
Collection.)

built from Burnells to Carlotta and from Eureka to Arcata.

The Noyo and Pudding Creek Railroad to the south was the steel that came to the Fort Bragg Redwood Company in the MacPherson and Wetherby mill and timber purchase. With this nucleus, C. R. Johnson started building the California Western along Pudding Creek and had reached Glenella, site of the Pudding Creek Lumber Company, by 1887. Ships brought in two locomotives and a discarded San Francisco street-car for rolling stock. Four years later the timber here was exhausted and to bring in more from the Noyo River area, the Union Lumber Company was formed with W. P. Plummer and C. L. White joining forces with C. R. Johnson. The railroad was then extended to the Noyo by means of a tun-

nel 1129 feet long, constructed mainly by imported Chinese. The local citizens were incensed, tried to drive them out but met the stonewall of Sheriff "Doc" Standley.

After the severe business recessions of the century end, which Union Lumber Company was able to weather, the firm purchased the Little Valley Lumber Company at Cleone, the deal including a short railroad using horse-drawn cars on strap rails. The mill and railroad were abandoned in 1904.

By that year Union Lumber Company had pushed its rail line eastward to Alpine and put into effect a regular passenger, mail and express service to that point, incorporating the following year as California Western Railroad and Navigation Company. By 1907 the road had reached

UNION'S LOCOMOTIVE NO. 2
Another view of engine shown
on page 146 after rolling down
from spectacular curves east of
Fort Bragg. (Photo Union Lumber Company Collection.)

Smith Creek, the next year Irmulco where Irvine and Muir had a mill, town and railroad. Within a few years passenger service was offered to Soda Springs.

Construction meanwhile had started west out of Willits, on the main north-south Northwestern Pacific. Tunnel No. 2, 790 feet long, was pushed through the mountain near Summit and the two tracks joined in November, 1911. Passenger service was now stepped up, a through sleeping car accommodations put into effect between Fort Bragg and Sausalito. This continued until 1929. The business peak had been reached and steam equipment began to be an expensive problem. A gasoline motor car replaced the day train, another was put into service in 1934, a third in 1941. Steam was further replaced by diesel-electric equipment and its use ceased completely in 1956.

In Humboldt County other railroad changes had taken place. In 1911 the Northwestern Pacific Railroad acquired the line of the Oregon & Eureka Railroad except the Little River Branch, of which they only acquired about a mile. The Hammond Lumber Company continued to use the rest for logging, running their log trains over the N. W. P. into Sonoma. Oreon & Eureka engine No. 6 was sold with the railroad and the crew also changed employers at the same time.

In 1932 the N.W.P. abandoned operations north of Korblex to Trinidad and the line between Korblex and Little River Junction was torn up. The Hammond Lumber Company bought the portion between Little River Junction and Trinidad for its logging operation. To restore connections with tidewater at Samoa the Hammond Lumber Company purchased the Little River Lumber Company which had a railroad between Little River and Humboldt Bay. This line, known as the Humboldt Northern had been purchased by the Little River Redwood Company in 1925 from Dolbeer and Carson Lumber Company which had used it to haul logs to tidewater for their mill at Eureka. The Little River Redwood mill at Crannell was closed down by Hammond, but the town became headquarters for the railroad.

CRUMMY AT BOYLE'S CAMP at Big River. Mendocino Lumber Company used this improvised car to haul crew to and from woods. (Photo Agriculture Extension, University of California.)

SPIDER-LEGGED TRESTLE (opposite) creeps across gulch. Piling for this logging trestle No. 112 on extension of the Eureka and Klamath Railroad was 80 to 130 feet long. (Photo California Redwood Association.) Above, Molino Timber Company engine and slab cars at Lola Prieta shipping station in Santa Cruz county. (Boeckenoogen Collection.) Below, California Western train at Fort Bragg about 1890. Charley Henningsen, engineer and shown also is John Ross. (Photo Union Lumber Company Collection.)

ANOTHER SONOMA CLASSIC (above) running mate to engine shown on page 135. (Photo California Redwood Association.) Below, left—Mendocino Lumber Company train at Hansen's Curve log dump. In vest, Charles Perkins, bull of the woods; next right, Frank Dutro; on his right, Tom Richards; on car back of engine, Tom McGraw; engineer is Joel Lilley; man with brake key, Jim Williams. Center, Caspar Lumber Company locomotives in early '80s. Right, moving cabins at Albion Lumber Company Winery Gulch camp in 1903. At extreme right, John Gunnar, at his left—George Escola. (Bottom three photos Escola Collection.)

"CASPAR WAS AN OLD MILL"

Alfred Quarnheim said this in 1906 when he was a boy starting to make lumber the hard way. From Caspar he went to Fort Bragg and Eureka, spent many years with Hammond and Holmes-Eureka Lumber Company, eventually becoming manager of the latter firm. His reminiscences here are of the days he was adrift in San Francisco and "sailed" into sawmilling.

"Jobs were hard to get in 1906 and it was still harder to find a place to live. I shipped out on a little gasoline motored schooner, *Pike County,* owned by Peterson and Crossley. We loaded 100 tons of wheat at San Francisco for Petaluma and had to sail up the slough at high tide—captain, engineer, myself and another young kid as deck hands. The first trip up the captain was drunk and made me to take the helm. I told him I did not know the slough. It was dark but he made me take over and went to bed.

"Everything went fine for a couple of hours, then the engineer told me we were stuck and to go to bed. In the morning I heard the captain holler and cuss and when I came up I saw the slough several feet from the boat. We were high and dry. If it had not been for the engineer, I believe the captain would have killed me but the engineer called him a drunken so and so and stood in front of me with a big wrench.

"We got off at high tide and into Petaluma at noon. There were six stevedores on the dock and only us two green deckhands to make up sling loads of 100 lb. sacks of wheat. The captain went ashore and sent us a sack of beer. I asked the stevedores if they wanted some and before we knew it two of them had come down in the

hold and we had the 100 tons out in no time. The captain was very surprised when he returned and asked the engineer 'How come?' When he got the answer he took all of us to town and treated us to a big dinner.

"I made a few more trips and that was the end of my sailor days. Hired out from Murray and Ready, an employment office, to go up near Sacramento to work on a railroad. Took the ferry up and landed at the supposed place—Freeport, I believe the name was. We asked a farmer where the railroad camp was and he told us there had been no railroad work for over two years. We got jobs on a farm near Walnut Grove, but that did not last long. Back in San Francisco, believe me, I got the money back from Murray and Ready.

"My friend and I had our experiences with the earthquake, one of which I remember vividly. The day after the quake, we were told to help clean up the rubbish in the street. Saloons and stores were being looted as they burned. One woman came out of a saloon carrying four bottles between the fingers of each hand. The Army captain told her to drop them and got a tongue lashing for it. We last saw her staring mournfully at the tops of the bottles. We had to be indoors by 8 p.m. and were a few minutes late one evening. Suddenly we were confronted with a gun and 'Halt! Who goes there?' The bags we carried were passed but that gun barrel looked awfully big and the memory got us in on time after that.

"We were rather down in the dumps by early fall and were sitting on a dock when a fellow came over and said 'Are you kids looking for a job?' We told him we sure were. He said to get our bags and come with him up to Caspar. 'Where is Caspar and what kind of work?' He told us it was in Mendocino County and we could work in a sawmill. Neither of us had ever heard of a mill or worked in one and told him so. He assured us it

EXCELSIOR REDWOOD ROLLING STOCK (Above and opposite) posed by photographer A. W. Ericson. (Pacific Lumber Company Collection.) Below, Engine No. 3 on Arcata and Mad River Railroad. Engineer, Grant Warren, brakeman, Freddie Pappine, conductor Fred Marquardt. (Photo Fred Marquardt Collection.)

was o.k. and if we were not satisfied he would take us back to San Francisco. We went aboard the steamer *Caspar* and slept in the hold. In the mill I learned to tail off the resaw—rather hard work. Ate in the cookhouse and slept in a rough cabin on a straw mattress.

"While in Caspar we went fishing off the rocks on Christmas eve. Did not notice the tide coming in and when we were ready to go home, we couldn't get off the rock—had to stay there all night in the rain. We were two tired, hungry and sleepy boys the next day. We had had all the ocean fishing we wanted but the following Sunday went up to Little River and caught a lot of nice trout. As we were sitting on the bank cleaning the fish we saw a black bear across the shallow river. We picked up our rods and fish and ran. We fell down a hill and heard the darndest squealing. We had landed in a pig pen—and thought the bear was after us.

"Caspar was an old mill. Lumber had to be hauled up an

e Mills, Humboldt Co, Cal.

incline to the yard. The schooners were anchored outside and lumber had to be transferred by wire cables.

"After the new year 1907 we quit our jobs and went to Fort Bragg and the Union Lumber Company. Mr. C. R. Johnson was president and a finer man never lived. Later I got to know his son Otis and his son Russell, who is now president. We worked here till spring, then went back to San Francisco. Still had no suitable jobs so we went up to Eureka and worked for Hammond Lumber Company. Spent 13½ years there and really got my lumber experience at Hammond. Worked 10 hours a day and to business college at night. Mr. W. R. McMillan was the superintendent and a tough Scotsman. However I have much to thank him for. Worked in the yard, then got on tallying which in those days was some different

than now. Doubt if there are many old time tallymen who could do what we had to. Lumber was taken down from high piles and loaded on three-wheel trucks—hauled by a horse. We usually had to tally for two gangs and keep track of board measure at the same time. For example when an offshore boat was loading green clear redwood, we had to tally it as it came from the sawmill and the balance from the yard. Say 6787 feet 1¼x8/24"x10/20'-5350 feet 1¾x8/24"·10/20' etc. and keep a running tally in your head so as not to overrun—was not an easy job, especially at night in the rain, under an umbrella, holding a coal oil lantern in the crook of your left arm, tally boom in the left hand, pencil in right. Try it some time.

"I remember one cold, stormy night we were loading

155

LANDING ABOVE MILLWOOD on Smith and Moore's Sierra operation. Chute brought logs to rollway and were railroaded down to Lower Mill. Photo below shows camp with bulls and horses yarding logs. (Photo Annand and Harold G. Schutt Collection.)

MAD RIVER MINIATURE at Glendale—one of the first railroading attempts in Humboldt County. (Ericson photo Pacific Lumber Company Collection.)

157

POLE ROAD POWER—Vertical boiler, reserve water tanks and concave wheels to fit poles sunk in the ground, were features of this 1872 locomotive which dragged logs between poles. (Photo Pacific Lumber Company Collection.)

one of the Hammond boats for their yard in Los Angeles. I was lucky. I had a pile of 4x4 No. 3 under the electric light and I could stand under the tramway. But Jim was tallying 3x4 in the next alley in the open. I heard some-one mumbling '12-18-20-10.' Then I saw Jim busily marking in his tally book. When I asked him what he doing he said: 'Tallying 3x4.' 'How can you do that here? Your men are down in the next alley.' He calmly said: 'I'm tallying by sound. Hear that? That's a 14 foot—that's 20x20. I can tell by the time it takes the 3x4 to come down from the pile!' I told him he'd better go to his men before the foreman got there. He did. We had a tall, thin Irishman—call him Dennis. He was loading slings for the steamer *Leggett* when something happened as it was being hoisted—lumber spilled all over the dock.

McMillan happened to be there and asked Dennis what the hell was the matter with him—did he not have any sense? Dennis eyed the super and said: 'Begorra, if Oy had any sense Oy would not be workin' on such a job.' Next day Dennis had a better one and McMillan often kidded him about it.

"One dark night we were loading the *City of Topeka* with shingles from piles on the dock. The dock planking was rather rotten with round holes here and there. The mate, kind of a dandy, came ashore to see how much more cargo there was and suddenly he disappeared through one of the holes, down into the mud and water. Jim thought it was funny—walked backwards laughing, and he too disappeared. We lowered a lantern and there they were looking startled at each other."

REDWOOD DOWN TO SEA

The solitary passenger on the little coastal steamer was restless. The ship had left San Francisco with general cargo and supplies for the lumber operation at Usal but a gale was roaring up the rocky headlands and she could not make port. Standing out from the mouth of the Noyo River she pitched and rolled like an egg in a boiling pot.

Leathery Capt. Bob Walvig knew there was no use trying to make things look better. The passenger knew as much about the "ole debbil sea" as he did and more about the California coast. So he pounded his knotty fists and pointed through the screen of spray to rocks off the port bow.

"North reef there and the channel there. In the Gilberts them natives wait for the proper wave and ride themselves over the reefs. We'll try it."

The passenger was anxious because he owned the Usal Lumber Company and around his waist was a money belt with the payroll he had secured in San Francisco. But he was a sturdy Scot, not often frightened for his own life and expecting others to look after theirs. Another thing—this was his ship, the 218 ton *Newsboy*, newly launched at Boole and Beaton's shipyard complete with engines, this very year 1888. He was Capt. Robert Dollar and through narrow eye slits he studied the low rocks slapped and smothered by the angry sea.

"You're master, Bob. Be sure you're right. You're no Gilbert Islander."

Like a balsa raft the *Newsboy* lifted to the crests and sloughed into the troughs with sudden, breathtaking dives. Walvig paired the wheel with his

LOADING SCHOONER AT GUALALA—Cable from high landing was secured to vessel's rigging and carriage block lowered lumber by gravity, pulled back by winch. (Photo California Redwood Association.)

BARKENTINE JANE L. STANFORD with cargo of redwood from Eureka. (Photo Worden Collection, Wells Fargo Bank History Room.)

helmsman and worked the schooner as near the barrier as he dared. He studied the water and measured the extremes and the ship squared to the reef when a thirty-foot breaker caught the stern. She rose like a whale and skated to the bottom not six feet beyond the rocks as the broken wave swept the deck. Robert Dollar hugged the pump handle and color returned to his face as the surf leveled.

"Great guns of war! Hoodlum Bob. No wonder they call you that. Bob Walvig—don't ever do that again. Not with my ship!"

Wet and grateful, Dollar telephoned his son Stanley in San Francisco, got a horse in Fort Bragg and paid off his men at Usal. Capt. Walvig went on to command the schooners *Cleone, Point Arena, Scotia* and *Quinault* while Capt. Dollar built a floating empire of steel with more than sixty ships. The *Newsboy* ended her career by colliding with the schooner *Wasp* on the Humboldt Bar in 1906 but she left her name on the charts of Fort Bragg harbor—Newsboy Channel.

The wooden ships of sail and steam were the lifelines of the redwod mills. Railroads were far beyond the mountains and wagon roads still a dream. Ships were the only contact lumber had with the market in San Francisco and the world.

Tall masts and taut lines marked every shipping point along the long redwood coast. There were no harbors worthy of the name and ships tied up at all manner of wharves or stood out of the Mendocino "dog holes," dismal little gashes in the rocky, inhospitable coastline—Little River, Greenwood, Bowen's Landing, Duncan's, Whitesboro and a dozen others.

The world has never seen anything as brash and inventive as this method of loading lumber. It was born of necessity and as crude as the early mills themselves. And at every step in the process, for every motion of handling the rough pieces of redwood and for every schooner anchoring, raw courage was the required element.

The Stewart and Hunter activity at Newport was illustrative of all the others. In 1878 the mill was on Mill Creek, a tributary of Ten Mile River, about five miles from the headlands. The logs were hauled out of the gulch by bulls, the lumber hauled by six-horse teams and hand-made wagons to the shipping point and piled at the chute head, seventy feet above the pounding surf.

This chute extended some eighty feet down and out over the rocks, the whole spidery contraption supported by a trestle and a type of A-frame. The apron or thirty-foot extreme lower section was suspended by cable from the A-frame and swayed in the perpetual wind like a grasshopper's antenae.

When one of the little single-ender lumber schooners was ready to chance an entrance to the channel between the rocks, she would signal ashore and prepare for the ordeal. This might take a few hours or a few days, depending on the weather and tide. Even after anchoring a ship might have to pull up and run for it, the loading operation halted until she got back in the dog hole again.

This anchoring process was at once an art, a specialized skill attempted only by the brave or the simple-minded. Having no shelter from the merciless wind and wild breakers, the shoals shallow and treacherous, a skipper would lay-to as

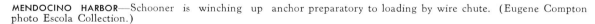

MENDOCINO HARBOR—Schooner is winching up anchor preparatory to loading by wire chute. (Eugene Compton photo Escola Collection.)

S. S. MELVILLE DOLLAR AT MENDOCINO in 1909 taking aboard 40,000 redwood ties for Mazatlan, Mexico. (Photo Union Lumber Company Collection.) (Center) Four small schooners waiting for cargo at Albion in 1897. (Bottom) *National City* at right anchored at Mendocino. (Photos Escola Collection.)

SEQUOIA LOADING AT CLEONE — "Wharf" it was called but lumber had to be sent out to ship by cable. *Sequoia* was only ship built at Fort Bragg. (Photo Escola Collection.)

close to the jagged rocks as possible and at just the right opportunity, order his pilot to break through. He had only one chance. If he missed it, the ship was aground and already being battered to pieces. Once into the hole, anchors would be dropped fore and aft and the ship winched either way to settle her as near the bottom end of the chute as possible. Sometimes spring lines were used to allow a twenty-five foot sweep under the chute.

Above the schooner's deck, riding the swaying apron was the clapperman, more catty and contemptuous of danger than any steel rigger on solid footing. At his signal the feeder sent boards rattling down the greased chute until it was full. Then if the ship was under him, he would release the first piece of lumber, the gap filling up. Then another and another as long as he had a deck to drop to. And so, in two or three days, the load was made up—seventy-five to one hundred thousand feet of redwood ties, shingles, fence posts and timbers on their uncertain way to market.

Only a few of the smallest steam schooners could safely negotiate these chancy jobs. They had engines aft, the foredecks clear for landing cargo. At Little River, Capt. Thomas H. Peterson built twenty of these schooners and sold them as fast as they came off the ways. At Navarro, Charles

Fletcher built several as did the yards at Oakland, Benecia, Vallejo.

More useful for the general trade were the schooners built especially for handling lumber from the wire chutes or wharves. They carried about two-fifths of their cargo in the hold, three-fifths as deck load and when loaded looked like floating lumber stacks.

A later type of steam schooner carried wooden beams, seventy to seventy-five feet in length which cut down labor in moving berth. Further fast handling was attained by winches with off-shore and inshore falls operated by one driver. Double-enders were later equipped with fast gear fore and aft, some ships built with runways through the midship house for long timbers stowed on deck.

Steam did not replace sail on the California coast for thirty years after the first engines converted the barkentines and schooners. Canvas was kept aloft for auxiliary power while the firemen stoked the boilers. But steam revolutionized lumber transportation by speeding up voyages and allowing ships to enter long forbidden channels and rivers. Yet the converted sailers were never the able profit makers the coast-built, steam lumber schooners were. Sailing vessels like the *Prentiss, Celia, Lakme* and *Laguna* went on regular redwood call while the black gang worked below but

163

SAMOA SHIPYARDS across the bay from Eureka where scores of wooden lumber schooners were built in the Humboldt heyday. (Photo Bancroft Library, University of California.) (Below), Robert Dollar's first ship, *Newsboy* built for his Usal Lumber Company. (Photo Union Lumber Company Collection.) (Bottom) Steamer *Marshfield* aground off mouth of Little River, unidentified ship standing by to give assistance. (Photo Escola Collection.)

owners looked forward to special lines, decks and stern ports for the more economical stowing of lumber cargo.

The *Daisy Mitchell* was one of these. She was built at Fairhaven, on Humboldt Bay, by Hans Bendixsen and had her engines and crew accomodations amidships which gave plenty of room for lumber below decks and on deck, easily placed with improved loading gear.

The Bendixsen yard was famous in its day. It was started by the Fay Brothers who, said the Humboldt Times on Dec. 14, 1878," "built the schooner Phoebe Fay, the steamer Pert which was taken to San Francisco and other vessels; they were followed by Captain Cousins who built the brig Hesperian, the barkentine Western Belle, the schooner May Queen, the steamer Humboldt now plying between this city and San Francisco and others.

"Bendixsen and McDonald followed to the business to which firm the former named gentleman preceeded. Captain H. E. Bendixsen is well known as one of the most successful shipbuilders on the coast. Some four years ago he removed his shipyard to Fairhaven. He has built the following named vessels.

"Schooners: Laura May, Luella, Fair Queen, Undine, Stella, Elvella, Mary, Marion, Atlanta, Aurora, John McCullock, Lovely, A. S. Fowler, Golden Gate, Humboldt, Venus, Albert and Edward, Lottie Collins, Pauline Collins, John N. In-

POINT ARENA WITH BIG DECKLOAD at Mendocino, *Seafoam* at anchor, beyond. (Eugene Compton photo Escola Collection.) (Below) Steam schooner *Brunswick*, of National Steamship Company, Union Lumber Company subsidiary. Built by Capt. A. M. Simpson at North Bend, Oregon, she was famous lumber carrier for Union and Hammond Lumber Company. (Photo Union Lumber Company Collection.

galls, Mary Swann, J. G. Wall, LaGironde, Vinne, Varro, Varco, Hinaerii, (these last four for the Tahiti Island trade), the Laura Pike, David and Ettie, Mary Buhne, Christine Steffens, Abbie, Lizzie, Madison, Martha W. Tuft, Gussie Klose, Morning Star, Excelsior, (three masts), San Buenaventura, Edward Parke, Maxin, Compeer, (three masts), Albert and Edward rebuilt, Orion and George R. Higgins.

"In addition to these he has built the barkentine Eureka, the brig Nautilus, the brig Paloma, steam-

ers Silva, Little Jones, Alta and rebuilt tug Mary Ann and several others. Names not available."

During the three years following the San Francisco quake and fire when the cry for lumber was one of desperation, more than thirty Pacific-type steam schooners slid down the ways in San Francisco Bay and other yards. Then there was a lull after which came the first World War and sudden demand. But with the war also came disaster for the wooden fleets. Steel was the new order and an era had passed.

"There was a lot of fun in these venerable Western ports," say the authors of Ships Of The Redwood Coast, Jack McNairn and Jerry MacMullen. "The two times of the year that are well remembered by inhabitants of the redwood coast are the Fourth of July and Christmas. At those times logging trains pulled into the coastal port terminals, such as Fort Bragg and Albion, from the logging camps with a full load of Finn loggers, ready for their hell-raising two weeks of fun and fight. A practice that was not exactly a credit to the pioneers was that of the saloon keepers when the loggers came to town. The mills paid off in checks from the San Francisco offices, each little more than a piece of paper to a rough and tough Finn, who wanted, not a drink, but a lot of drinks after six months in the woods. Once paid off, the boys would go to the saloons, where their checks were cashed at from five to ten per cent discount, and then they would just stay at the bar and spend their money until it was all gone.

"Old timers pass along the story of Greenwood in the early days when the company owned the only saloon and hotel in town. Here they paid off the loggers in the same manner, by check from the

San Francisco office, and here they discounted the checks at the company saloon or hotel, charging five or ten per cent, wich is a tidy discount from a $500 check, the average logger's pay for six months. Then they accepted the entire business of the loggers, in drinks and quarters. And when the boys had used up all but a few dollars of their hard-earned money they would board the logging train and go back to the woods, to work and await another hell-raising time six months later.

"But not all the towns were the same. Standish and Hickey at Albion frowned upon saloons in the mill town and dispensed no liquor on their property. But along came a young fellow, one Bob Mills, who bought a parcel of property across the river from the mill town and made a fortune in the saloon he operated. Discounting, the usual practice here too, added to his income. There were those who said he bought a barrel of whiskey and sold ten barrels of drinks from it; but then perhaps it had to be cut to that extreme to make it palatable.

"Little River ,too, is somewhat in the same situation as the other abandoned mill towns. There is no trace of the mill, the wharves and chutes are gone, and most of the population that made up the town has moved away. The cove that once harbored lumber schooners now caters to abalone fishermen. A few old homes remain. Silas Coombs' mansion, turned into an inn bv his granddaughter, has been the haven of many celebrities, including the Governor of California. Trails that at one time felt the hooves of oxen hauling logs to the mill now feel the pad of fishermen as they beat their way to pools of Little River. Hollywood took over the locality during the filming of "Frenchman's Creek", a story based on the English coast; their movie-made sailing ship anchored in the cove so long used by schooners of Silas Coombs. Charles Van Damme, a lumberman and San Francisco fer-

BOUND FOR DOWN UNDER — American schooner *Irene* being loaded at Noyo wharf Sept. 9, 1915 with 900 thousand feet of redwood for Adelaide, N. S. W. Carriage is bringing men back to wharf. (Photo Union Lumber Company Collection.) (Below) Steam schooner *Alcatraz* wrecked in 1917 on Mile Rock one clear, moonlit night. (Photo Union Lumber Company Collection.) (Right) Steam schooner *Tamalpais.* (Photo Worden Collection, Wells Fargo Bank History Room.)

ryboat operator, purchased the land surrounding Little River, the place of his birth, and upon his death his wife donated it to the State as a park.

"Mendocino City, formerly Big River, with a population of less than eight hundred in 1940, is practically a ghost town, in comparison to its former size. The closing of the large mill here in 1931 virtually killed the town. Greenwood, just south of Albion, is another community which has gone the way of all abandoned mill towns, but she has had a roaring good time during the century of her existence. The Greenwood brothers — William, Boggs, James and Britt—settled in this area in the late 1840's, and they say it was Britt who led rescuers to find the survivors of the ill-fated Donner Party. Greenwood became prominent in the late 1890's when Lorenzo E. White built a large mill here...

"For forty years the L. E. White mills operated here, White becoming a power along the coast. It was only fourteen hours from Greenwood to San Francisco by steam schooner and vessels loaded one day could leave at dawn the next morning and be in San Francisco unloading by the same evening. White operated two of the early-day converted sailing-steam schooners, the *Alcatraz* and the *Alcazar* and many two-masted schooners. An incline ran from the 150-foot cliffs down to the most easterly of a cluster of three large rocks, from which a wire cable led for loading and discharging. Vessels rode to moorings on the north side of the outermost rock, called Casket Rock. At night the glare from the electric lights was conspicuous at sea, but there have been no electric lights since the mill shut down in 1929.

"Greenwood went the same way as all of the lumber towns when the timber was available only at distances, making it unprofitable to ship to the mill and down the coast by steamer. In the 1930's the Greenwood mill was dismantled and the machinery sold. The wire-cable landing and trestle have fallen apart and the mill has all but disappeared, just a few piles remaining to remind one of a once prominent enterprise."

TANKS FOR TASMANIA (Opposite)—Redwood tanks being loaded at Oakland on the *Sibyl Martin*. (Photo Bancroft Library, University of California. (Below, left) Steamer *Jeanne* stranded near Point Arena in 1900. (Center) Wreck of S. S. *Pomona* off Mendocino coast. (Right) Steamer *Santa Barbara* under salvage tow. (Photos Worden Collection, Wells Fargo Bank History Room.)

Wreck of S. S. Pomona of Mendocino.

WAVES BEAT FUNERAL MARCH as moonlight sheds pattern of beauty on remains of a gallant lumber carrier.

NOYO WAS UNION'S FIRST steamer, shown here loading at Noyo harbor. Built at Bendixsen, Humboldt Bay, in 1888, she hit Bull Rock off Albion while in tow to San Francisco and sank off Point Reyes Feb. 26, 1918. (Photo Union Lumber Company Collection.)

WRECK AT KENT'S POINT — Schooner *Bobolink* in deep trouble on the rocks March 24, 1898. The cook was washed
overboard, lumber salvaged before ship broke up.

SCHOONERS J. M. GRIFFITHS (left), *Nahomis* (right) and steamship *No Wood* (below) all heavily loaded with redwood. (Photos Worden Collection, Wells Fargo Bank History Room.)

SCHOONER NORTH FORK leaving Hammond Lumber Company docks at Samoa, Eureka Bay, in 1915. (Photo Ag. Ext. Collection, University of Calif.)

SCHOONER ELECTRA riding high into Medocino Bay to load lumber. (Photo Escola Collection.)

STRANGER FROM ABROAD Five masted German bark *Kobenhaven* loading at Oakland. (Photo Bancroft Library, University of California.)